EAST OF LIBERAL

THE DREAMSEEKER MEMOIR SERIES

On an occasional and highly selective basis, books in the DreamSeeker Memoir Series, intended to make available fine memoirs by writers whose works at least implicitly arise from or engage Anabaptist-related contexts, themes, or interests, are published by Cascadia Publishing House LLC under the DreamSeeker Books imprint. Cascadia oversees content of these novels or story collections in collaboration with DreamSeeker Memoir Series Editor Jeff Gundy.

1. Yoder School
   By Phyllis Miller Swartz, 2019
2. East of Liberal: Notes on the Land
   By Raylene Hinz-Penner, 2022
3. New Moves: A Theological Odyssey
   By J. Denny Weaver, 2023

EAST OF LIBERAL

Notes on the Land

A memoir by

Raylene Hinz-Penner

<u>DreamSeeker Memoir Series, Volume 2</u>

DreamSeeker Books
TELFORD, PENNSYLVANIA

an imprint of
Cascadia Publishing House LLC

Cascadia Publishing House orders, information, reprint permissions:
contact@CascadiaPublishingHouse.com
1-215-723-9125
126 Klingerman Road, Telford PA 18969
www.CascadiaPublishingHouse.com

East of Liberal

DreamSeeker Books is an imprint of Cascadia Publishing House LLC
ISBN 13: 978-1-68027-022-8
Book design by Cascadia Publishing House
Cover design by Gwen M. Stamm based on photo of aerial map provided by Raylene Hinz-Penner and picturing the author's familial homestead, with the specific acreage outlined in maroon.

Library of Congress Cataloguing-in-Publication Data

Names: Hinz-Penner, Raylene, author.
Title: East of Liberal : notes on the land / a memoir by Raylene Hinz-Penner.
Description: Telford, Pennsylvania : DreamSeeker Books, [2022] | Series: Dreamseeker memoir series ; volume 2 | Includes bibliographical references. | Summary: "Through elements of memoir, history, and philosophy of land use, a lover of her Mennonite farm childhood looks critically at farming's impact on the land, comparing settler values of land ownership to those of first peoples who see themselves as owned by the sacred homeland"-- Provided by publisher.
Identifiers: LCCN 2022038028 | ISBN 9781680270228 (trade paperback)
Subjects: LCSH: Hinz-Penner, Raylene--Childhood and youth | Mennonite farmers--Kansas--Liberal--Biography | Farm life--Kansas--Liberal. | Liberal (Kan.)--Biography
Classification: LCC S521.5.K2 H56 2022 | DDC 630.9781/735--dc23/eng/20221019
LC record available at https://lccn.loc.gov/2022038028

28 27 26 25 24 23 22 10 9 8 7 6 5 4 3 2 1

To
my parents,
who spent their lives
working on and loving the land

Contents

I have read that the Apaches believe land makes people live right. Can that be true? The Apaches also honor place as the origin of story. I know that to be true. It is said they begin and end every story, "It happened at. . . ."

For us, it happened three miles east of Liberal in the corner of Seward County, Kansas bordering Beaver County, the Oklahoma Panhandle once known as "No Man's Land." My parents, a young and eager post-World War II couple, were looking for a place to farm and considered themselves unaccountably lucky in 1950 to have found a half-section of sand left mostly untended since its topsoil had blown away during the 1930s catastrophe known as The Dust Bowl.

The children of generations of Mennonite farmers, my parents set about bringing the land back into productivity. Realizing almost immediately that the dry land would not sustain them, they accumulated a herd of Holstein milk cows and managed the dairy together for a quarter century. Purportedly, Menno Simons, the founder of our faith, was the child of a dairy farmer. So was Georgia O'Keeffe. So am I.

PREFACE

Early in 2019 I found myself among a group of travelers retracing the first stops on the "Ruta de Cortes" 500 years after the Spanish conquistador's landing on a Good Friday near what is today Veracruz harbor in the Gulf of Mexico. Led by an archaeologist friend, our trip was not a celebration of Cortes' conquest of Mexico but rather a chance to explore archaeological sites still evident today on Cortes' route. We had come to see what remained of those who had been there before Cortes' arrival. Beside the long, earth-embracing horizontal branches of the ceiba tree in Antiguato—to which Cortes purportedly lashed his ships, a tree still rooted in the underworld according to the beliefs of the people who lived there, I had a kind of epiphany.

I had only recently participated in a heritage tour to Poland to visit the remnants of villages where generations of my people had once lived. My European farming ancestors' journeys from the Netherlands to Austria, Prussia, Russia, and eventually, to the United States had occurred in the same 500-year span since Cortes landed at Veracruz. What would I find if I examined the 500 years before my parents came to the land east of Liberal in 1950?

On the Totonac site at Zempoala, where people had lived for 300 years before Cortes' arrival, I recognized how impossible it would have been for those who greeted Cortes to foresee how forced slavery, smallpox, and war would decimate their mighty populations. We stood among the ruins of one of the twelve compounds of what was once Zempoala, nestled into the hills and valleys for protection from the elements and their enemies. Its natural beauty stands in stark contrast to the sites built by Spanish intruders, for example, the Fort of San Juan de Ulua in Veracruz harbor, notable for its massive prison walls. A chill rippled down my backbone high above the shore at Quiahuitzlan, the ruin overlooking the Gulf where the Totonac people would have watched the incoming Spanish ships. We ate lunch on the beach at beautiful Playa Villa Rica where Cortes showed off his horses and cannons to stun the locals into submission. The history I know is a chronicle of battles, a count of the dead from opposing sides, a measure of land taken, evidence of power wielded over the newly enslaved.

Just over two decades after Cortes landed at Veracruz along the Gulf of Mexico, Francisco Vasquez de Coronado left his post as governor in northern Mexico with the same gold lust that had driven Cortes to the shores of "New Spain." Coronado began a march north to search for the Seven Cities of Cibola. Crossing the land we call Kansas today, Coronado was looking for a Quiviran village whose inhabitants, he had been told, drank from jugs of gold. The Indigenous people he found, believed to be predecessors of the Wichita people, amazed him with their health, strength and height, but they had no gold. Disgusted and convinced he had been purposely misled, Coronado killed the Turk who had misinformed him and headed back south to Mexico.

Maps which plot Coronado's trek into Kansas show his path veering northeast somewhere near the farm I grew up on, three miles east of Liberal. Indeed, in 1979 The Smithsonian Museum authenticated an intricate set of buckles and a bridle bit found northeast of Liberal in Horsethief Cave in the 1920s, Spanish ironwork dating from the time of Coronado's expedition through the area in 1541.[1] Smoke on the roof of Horsethief Cave showed evidence of use even earlier by local peoples. Coronado's expedition would have fanned out by day and camped near the Cimarron River by night, possibly crossing directly over the land my parents would farm 400 years later. Today, the Spanish horseman's accessories are proudly displayed in the entrance to the Coronado Museum on the eastern edge of Liberal, not far from where our rural mailbox once was.

Shortly after Coronado crossed Kansas looking for Quivira, another European named Menno Simons, the priest who would leave the Catholic Church and become an Anabaptist revolutionary, visited the earliest Mennonite church in Gdansk in 1549. This priest is my religious ancestor and the instigator of my family's migrations. My farming ancestors began moving 500 years ago, from the Netherlands to Mennonite farming villages across Prussia including Gdansk, an area to which Mennonites fled persecution. This book traces the migrations which eventually landed my family east of Liberal; these notes document my own wrestling with peoplehood in my tradition. I had always thought of us as immigrants helping to build a country where freedom could prevail, not colonizers or settlers. In fact, I had mythologized my heritage as the legacy of a people purified by the fires of martyrdom, simple hardworking farmers looking for refuge from the state. The story is not that simple.

* * *

Numerous persons contributed to my engagements with the story's many layers. I wish to thank Jeff Gundy for his careful reading and valuable suggestions and Michael A. King for his willingness to honor my vision for a book of notes and shifting genres. I acknowledge my mother and sister for helping me to retrieve memories. I thank the fellow writers who have read pieces of this text and said, "Keep going." Thank you, Julia Kasdorf, for advising me to "write the change." Most of all, my husband Doug who reads upon request, listens patiently to what I intend, and invariably offers the questions, clarifications, or affirmations I need.

—Raylene Hinz-Penner
North Newton, Kansas

1949

I was born in the first month of the year, the month named for Janus, the Romans' two-faced gate god who could look back into the past and forward into the future. Appropriately for this story, Janus was also the Roman god associated with the cultivation of fields.

The day after I was born, Harry Truman was inaugurated to serve a second term. Liberal's Southwest Daily Times*—the local newspaper no doubt lying folded on a table in the Epworth Hospital lobby where my father would have eagerly pounced on it—announced Truman's theme of "Peace, Plenty and Freedom" for the world: "Our aim should be to help the free peoples of the world, through their own efforts, to produce more food, more clothing, more materials for housing, and more mechanical power to lighten their burdens. . . ."*

Alongside the map of Truman's inaugural route in the Times *was a report on harvest labor, just released from the capital city of Topeka, announcing the intent of the Kansas State Employment Service to improve the ability of Kansas farmers to get their wheat cut in 1949. During the 1948 harvest, 58,546 Kansas farm-owned and 3,788 out-of-state custom combines harvested just short of 15 million acres of wheat on Kansas farms.*

A year later my parents would join that throng of Kansas wheat farmers. One year to the day after my January birth my parents moved onto the land east of Liberal. 1949 had been a record-breaking year of rainfall, with western Kansas receiving 31.14 inches, nearly forty percent more than usual. Southwest Kansas had been mentioned in the Kiplinger Letter as a favored area for expansion. My parents were thrilled to get their chance to farm during the expansionism of 1950, when most land in temperate climates had already been claimed for farming.

1950

Chapter One

A Coyote Hunters' Shack

My father and his eldest brother are throwing bottles—deep brown and clear-mottled and cobalt blue—onto the back of a faded old green pickup truck parked three miles east of Liberal on a sandy road which will later be known as Pine Street. There are no pines on this forsaken piece of property. The two men stand next to a coyote hunters' shack my father has agreed to live in. Whiskey bottles, pork and bean cans, jagged and sharp, knife-opened tins. You learn something from what people throw out and where they throw it.

As the story came down to me, ten people had been living in the four tiny rooms of the shanty, coyote skins stretched over the walls. My mother and father would soon find the remnants of a "moonshine still" half-buried in the scrub trees west of the house that had overtaken the farmstead during the years following the 1933 tornado.

Inside, the lath walls were barren of plaster. In fact, the walls had been pulled apart by tornadic winds and left unrepaired. Daylight streamed through the open fissures, as did creatures other than human. Doc Blackmer owned the place and wanted to see it cleaned up. But my father could not yet bring his young wife and baby daughter to this hovel, though he knew he must get them out of the tiny, lonely apartment my mother despised fifteen miles away in the little Oklahoma Panhandle town of Turpin.

Mama's older sister came from the sod house built by her homesteading in-laws in Texas County near the Texas/Oklahoma border thirty miles south of the farm east of Liberal, and declared the place uninhabitable. But when Mama and Daddy insisted that they would live there, Aunt Esther returned with a dozen laying hens to help them begin their lives on the land. There was no electricity, because the Rural Electric Association had not yet run power to this desolate farm. My older cousin remembers a bulb dangling from a long cord in the middle of the room, but that must have been at a later birthday party, after my mother could dispense with the oil lamp she carried room to room in 1950.

The two brothers stand together in the garbage pile beside the front door. My father's eldest brother, in fact, had "lured" him from central Oklahoma where Daddy grew up and married, from the bookkeeping and accounting jobs in Clinton he and my mother abandoned to farm and "be their own bosses." Uncle Elmer was certain my parents could find land in the Mennonite farm community in the Oklahoma Panhandle near Turpin, Oklahoma, just across the border from Liberal. Land abandoned during the 1930s could provide my father the opportunity he wanted. But months had turned to years, and my parents had not found land. Instead, my father had worked as a farm hand, gone on a custom harvest crew to cut wheat as far north as Montana up to the Canadian border. Now they had a baby girl; Daddy was desperate by January 1950. A coyote hunters' shack looked good.

The land east of Liberal was owned by my uncle's physician, a prominent doctor, banker and horseman from Hooker, Oklahoma, twenty miles west. Doc Blackmer had either bought up or foreclosed on Dust Bowl land during the years of hard times. I expect my father and his brother laughed at themselves and their plight as they took up their roles as garbage pickers, rescuing the shack where my father intended to live.

They would have joked in Plaut Dietsch, the language they used for "making *schput*" (poking fun)—the language of lesser stature than the proper High German their grandfather spoke to them, church service language. A bystander would have heard the phrase we children often overheard muttered by adult Low German speakers as they shook their heads about a crazy, terrible, or unbelievable situation: *Gans fe'rekjt*. "Crazy." We didn't learn the language, but we knew the bawdy or off-color phrases we weren't supposed to hear. No matter how rundown and forlorn this place might be, my parents needed to move, quit paying rent, and get to work. But with what seed money?

Our place east of Liberal was not a part of the Turpin Mennonite community where my parents attended church at Friedensfeld ("field of peace"). Those Mennonites had weathered the Dust Bowl and stayed on their land. This was abandoned, sandy soil long left to blow. The Osage Orange hedge posts used for fencing were half buried in sand. Of the few outbuildings, only one seemed useable: a sizeable tin shed left from another time when someone had farmed and dreamed big here, evidence the large Chinese elm trees which rimmed the small house. No toilet. That fact surely annoyed them both. What had these people used? The first thing my father had to do was build a toilet. Eventually, he secured a good WPA-built outdoor facility, the clean and comforting two-seater I grew up using, but for now he had to tear down one of the ramshackle buildings and use the lumber to build a toilet.

The brothers, one the oldest in a family of seven, my father the middle child, eventually lived ten miles apart across the state line and shared both good and hard times. My Uncle Elmer was established on a farm near the church in the center of the community. Daddy, who had gone as a noncombatant into the U.S. Army at age eighteen after his father's premature death, had seen the world by the family's standards, and now had come to settle near his older brother.

The brothers' bloody link and subsequent estrangement came years later when our family was more prosperous. Once again, they were working together, this time cutting ensilage to fill the silos for winter feed for my father's dairy cattle. The juicy seven-feet tall fodder is sweet as sugar cane as it is cut, chopped, and hauled to the silos, two new ones by now, tall and straight behind the dairy barn. But the belt on the ensilage cutter needs to be greased. I imagine my father grabbing the bucket and brush to daub it—though they both know the elder brother to be the mechanic. In my mind—for no one wanted to talk about it, Uncle Elmer takes the bucket. "Here, let me do that." Because it is dangerous? Because he mistrusts his younger brother's ability? In a moment of throwback to when he was younger and his parents would say, "Watch out for your baby brother?"

The belt grinds off Uncle Elmer's hand six inches above his wrist. He must have howled in shock and then pain, knowing immediately how this stubbed arm would change his life. What I know is that my father roared past our house with the vehicle horn blaring and headed west into the Epworth Hospital emergency room with rags wrapped as a tourniquet to try to stem the flow of blood from Uncle Elmer's arm. The brothers' uncle, an earlier No Man's Land settler, came back to get the severed hand lying in the sandy field after his nephew's life had been saved, but of course, it could not be re-attached. The eldest brother of my father's family would stand ever afterward, our family's version of Ahab, this one holding his stump. Seeing Uncle Elmer was always a terrible reminder for us that it had happened on our farm east of Liberal.

PART I

WINTER SOLSTICE

The seasonal calendar of the Ancestral Puebloans outlines practices, rituals, and ceremonies according to the cycle of the seasons. Spring and summer: farming and plant gathering; fall and winter: hunting. Ceremonies, rituals, and dances occurred between the solstices and equinoxes: ritual meditations on seasonal work. The Anasazi annual cycle is organic, earthbound, recognizable to agricultural peoples across time who know what it means to prepare the head and heart for work and worship according to the seasons, familiar if one knows the land or farms the land.

I faithfully circle the Algonquin full moons on my calendar, follow the earth's changes with each month's full moon, think about how earlier peoples on this land named them: Wolf Moon for the howling of wolves in the cold still air of January; Snow or Hunger Moon for deep snow and the difficulty of finding food in February; Worm Moon for melting snow and the March last moon of winter; Pink Moon for new growth of grass pink or wild phlox; Flower Moon of May for flowers in full bloom and corn planting time; Strawberry Moon of June, a universal name for the full moon for all tribes wherever they were; Buck Moon of July for the velvety new antler growth on a young buck; Sturgeon Moon for August's plenitude of sturgeon in the Great Lakes where so many tribes lived for so long; the Harvest Moon of September, the Hunter's Moon of October, the Beaver Moon of November when beavers are preparing for winter: set the beaver traps for their fur will be needed in the upcoming cold. And the Cold Moon of December, the Long Night Moon.

A farmer remarks on the lack of light at winter solstice, the shortest day of the year. My parents recall their longing for more daylight when they moved to the farm and had so much work to do to make it habitable (and no electricity to prolong the work day). My parents dreaded going to the dairy barn at 5:30 p.m., nearly dark in December and January. After the winter solstice they begin to envision each day a bit longer, a bit lighter. Longer days mean thinking forward, setting goals, planning for a new year. New Year's cookies on New Year's Day. Like bears packing on weight for the long hibernation, we feasted to celebrate a new beginning.

Traditional Mennonites had "Watchnight" services on New Year's Eve. Vestiges of that tradition remained during my childhood as we were instructed to "watch and wait" for Christ's coming, perhaps in the upcoming year. In Russia, German Mennonites gave their Russian neighbors New Year's cookies when they came and sang for them on New Year's Day. In this tradition the fritters symbolized affluence, good crops, and good wishes to a neighbor for prosperity and a happy new year.

Perhaps the immigrant Mennonites in Russia were also buying the good will of their neighbors. The Russian Germans had a nursery rhyme that my mother's mother recited in Low German:

Eck sach den Shornsteen Roacke
Eck visst voll vaut ye moachke.
Ye backte Niejoash Koake.
Yave ye me eane
Dann bliev eck stoane
Yave ye me twea
Dann fang eck aun to goane
Yave ye me drea, fea, feef toaglick
Donn vensch eck you daut gaunse Himmelrick.

The English translation:
I saw your chimney smoking.

I knew what you were making.
You were baking New Year's Cookies.
Give me one—I stand still.
Give me two—I start walking.
Give me three, four, five at once,
Then I wish you the kingdom of Heaven.

As an adult, I have always understood that I should call my neighbors on New Year's Day and invite them to come sample the fritters hot out of the oil where I fry them—and invite them to take some home. Entirely unconscious of the fact that I was reenacting an old Mennonite tradition that was supposed to bring good luck or the good will of neighbors who had never heard of "*porzelchen*," it just seemed a right way to begin a new year.

The 1950s

Chapter Two

Alfalfa

Alfalfa: [Sp.< Ar. al-fasfasah, the best fodder] a deep-rooted plant of the legume family, with small divided leaves, purple cloverlike flowers, and spiral pods, used extensively in the U.S. for fodder, pasture, and as a cover crop. . . .

As long as I can remember, there were alfalfa bales. Four feet long, less than two feet wide, a couple of feet deep, brick-shaped, these bales were key to our existence as dairy farmers. Typically, the semi-truck load of hay would arrive at our farm in the night . . . from Colorado, or somewhere Daddy could make a good deal on the sweet-smelling green bales.

I remember being awakened by the commotion on our farm at 1:30 a.m., the downshifting of gears on the truck, the heavy wheezing brakes announcing the arrival of a load of alfalfa. I smelled the truck's exhaust fumes as I lay listening to its loud engine idle in the yard, red taillights glowing against our white dairy barn. The Holsteins went crazy bawling at the smell of freshly baled alfalfa. They seemed ready to riot, head-butting, racing—never mind their swinging udders—drooling as they reached through the corral to get nearer the alfalfa. They were not that hungry; they had been fed. They were that excited.

I never helped stack the bales, but I sometimes went to a window to watch the night arrival of alfalfa. Daddy pulled on his overalls and hustled outside to show the men where to build the stack northeast of the dairy barn, next to the long concrete feeding trough. Into that long trough he would walk at feeding time every evening, carrying bale after bale, cutting with his pliers the baling wire or twine firmly holding each bale. Daddy broke the bales as one would break communion bread, parceling them into smaller four-inch leaves which he splayed along the forty-foot trough where the cows were lined, side by side, heads down into the trough.

Getting hay stacked right was as important to a farmer as laying brick is to a brick layer. The stack needed to stand strong for months, through all kinds of weather, even while one end was being pulled off for daily feedings. Daddy tolerated no broken bales; the twine must be tied carefully, taut, to keep the bales tight and uniform. Misshapen bales created sag and the eventual collapse of a stack which needed to survive the winter. Was the alfalfa green? Well-baled? Securely tied? Had it been rained on in the field? Faded?—lying too long in the field under a drying sun? All considerations in the purchase of good alfalfa.

When that alfalfa stack was completed, my father felt secure. He could feed his dairy herd. All year he scoured newspaper ads, *Hoard's Dairyman*, flyers at the Standard Supply store, in search of hay for his cows. Who do you trust not to slip into the mix some moldy or weed-filled bales? I remember how he opened a bale to examine its texture, how he inhaled deeply the sweet aroma like a good brandy—though Daddy would not have known a good brandy!—how he held a thin square from the middle of a bale to his nose with the chicken-leg colored leather glove he always wore.

The alfalfa stack was my cathedral on the plains. Stair steps created as my father steadily pulled out the evening's count of bales and spread them for his herd, changed day by day my stair-step access to the top of the stack. Long legs like mine could ascend to the top of the alfalfa stack in something that felt like a Super Woman's leap, and from that perch I could watch the southern horizon: miles of flatlands. There I sat on my throne in the evening to think, looking out over Oklahoma—only three miles away as the crow flies—the state of my people, my ancestry. There I watched the heat waves massage and mirage the fields in the distance while I watched for the upstart whirlwind's brief scuttle into extinction. The Plaut Dietsch word for these whirlwinds sounded like a spin or a twirl on the tongue, a rough, deep-from-the-earth start from nowhere in its harsh consonant takeoff : *Kjriesel*, pronounced something like "kud-easel" with a gutteral "ch."

I watched the tumbleweeds roll across the road into the fence line where they caught and threatened to pull down the barbed wire as the seasons changed. I listened to the meadowlarks on a nearby post, heard the mockingbirds in the trees, watched the long-legged killdeer running the cow lot among the cattle, and followed the scissor-tailed flycatchers' swoops from the highline wire. I oversaw the cattle grinding away at the sweet alfalfa, their long, drooling saliva streams specked with the grain they were fed during milking. From my vantage point on the alfalfa stack I watched the neighbors work their fields as the wheat heads ripened and lost themselves to harvest. I surveyed the darkening gold stubble through which I would run errands into the wheat field, scratching to blood my bare legs.

I also watched for the first star and wondered about the first people, those who farmed here first, those who had left the arrow heads, those who walked this land before those who left the arrow heads. I listened for the coyotes' night yips and howls. And I watched our trees grow, our garden grow, our baby calves grow, from my perch on the alfalfa stack. I watched my own feet grow stretched before me on this alfalfa stack. Alf-Alf-A. I loved to spell it fast, a strange repetitious spelling, but I was not curious then about the Spanish-Arabic origins of the word. The word spelled sweet, sweet time to me, the slow time of my childhood years. The natural time that is the working of the universe.

On Work

I set out to tell this story of life on the land. As I moved through generational time I began to recognize the mythologies created by story, the ways our beliefs are shaped in community. I am only beginning to recognize what it means to be a Mennonite settler.

First, I have had to re-examine my attitude about work, doing, production, farming. In my experience in Mennonite communities, hard work is glorified. I was taught to value another's work ethic above all else. Thus, I am in awe of my parents' hard work in the reclamation of our farm east of Liberal; I have always told our story as our variation of the American success story. Though they were not first-generation immigrants, I saw my parents as being cut from that mold, having that "outsider" first-generation prove-yourself-worthy work ethic found in so many first-generation immigrants. They came to the land east of Liberal with a ten-dollar bill and made a life on land no one wanted. In many ways they were heroic. But I am beginning to recognize our pride in the proverbial Mennonite work ethic, the White Anglo Saxon Protestant shield that has protected us for generations.

Today I see also the underside of the mythology created around hard work as salvation, that central Mennonite tenet that comes down through generations of farming Mennonites designating work as sacred. Does our strong belief in the value of hard work, the key to successful farming, also allow us to privilege ourselves? To say that we have earned, even deserve, rights to the land? We have always considered land unused, even unoccupied, unless it is being "worked"—translate "cultivated," or "farmed" ("subdued?"), one of the ways our ancestors made false assumptions about the land they found when they came to this country to farm. My ancestral family settled among the Cheyenne and Arapaho allotments in central Oklahoma and sometimes resented their need to rent "Indian land" which wasn't being farmed, believing that their Native neighbors "didn't care for" the land they had received as allotments; farmers would have used the land more productively. They saw unfarmed land as wasted.

As a child we sang the hymn, "Work for the Night Is Coming" based on John 9:4; that song suggested that we must work now, for the night is coming when we cannot work. There is a frenzied notion of haste and desperation in that phrasing. The hymn instructs us to give "every flying minute something to keep in store." Is that notion indicative of our sense that our work is about production, storing up, even getting all we can to pass on an inheritance? Building bigger barns? Commodification? Farmers work to have land to give to their children. One reason my parents had to work so hard on the land east of Liberal is that they had inherited no land; both their parents had lost their land in central Oklahoma during the Great Depression.

I have begun to wonder too about my ancestors' need to farm. Is it in our DNA? Why would my parents leave the jobs they had in central Oklahoma where they grew up to go west and seek their own land? Do our people have an innate need to own land? Or were they simply seeking community? Autonomy? Resisting post-World War II American norms or embracing them? Sometimes I question what our work ethic has done to encourage our pride, our demand for self-determination and self-sufficiency. So closely tied with economic success and land ownership, I wonder, has it corrupted us?

100 million years ago

Chapter Three

Inland Sea

Thousands of years before all those evenings I sat atop the alfalfa stack, hundreds of thousands of years before the alfalfa stack or the sandy soil on which it stood, the Gulf of Mexico covered the center of this continent; water had not yet deposited the variety of soils which would one day become our farm. This was not so difficult to imagine on a hot summer afternoon. Staring south into Oklahoma from my perch on the alfalfa stack, I saw an almost liquid waving: the wheat stalks three feet high, bearded heads moved in a rhythmic golden sway, creating a liquid mirage in blue air. Boundaries were dissolved in the summer heat, and everything became permeable, intermixed—the land with the air. An eye squint merged the land and sky into sea, dissolved the crazy quilt, human-drawn boundaries—state lines, county lines, townships, and roads.

Fossil hunters have extracted sea monsters from this area. As citizens of Kansas we were asked to vote to select the state fossil—should it be the Xiphactinus or Mosasaur? Still encased in chalk that the voting public helped to painstakingly clean lay the seventeen-foot Xiphactinus. Indeed, Kansas has some of the best specimens in the world from the Cretaceous Sea. The mosasaur, a sea lizard, might have grown to be as large as 45 feet long and weigh seven tons.

The Western Interior Sea that once divided North America into east and west land masses, stretching through the center of this continent from the Arctic to the Gulf of Mexico, was a shallow sea, only about 600 feet deep. Formed when the sea level was at its highest, in the area where Kansas is now located, the sediments were deposited at a rate which would ultimately produce about one inch of hardened chalk for every 700 years of time.[2] This sea covered Kansas most of the last 70 million years of the Age of Dinosaurs.

The Cretaceous Period lasted from 144 million years ago until 65 million years ago. We are told that "drainage from the older North American continent to the east and the mountains rising on the new land to the west carried vast amounts of soil, sand, gravel into this seaway creating intermixed layers of sandstone, shale and mudstones along the shorelines. In the clear waters at the center of the seaway, calcium carbonate shells of untold billions of microscopic creatures produced thick layers of limestone and chalk."[3] Here is another boundary, the flat muddy bottom of the Inland Sea. Maybe 600 feet deep is a shallow sea; I measure it at about the depth of ten silos stacked end to end like the one I sat alongside on the alfalfa bales.

I somehow knew intuitively about prehistoric times and this inland sea as a child, long before I studied it as the Inland Sea because of my Great Uncle Albert's prescient stories. You could see he knew mysterious things in his watery blue eyes, glassy and milky like old marbles that roll around in a drawer—the rarer big ones. My father never believed his Uncle Albert was a great farmer, but he was a great teller of tall tales. We never quite knew if a story was imaginary, and the narrative's movement from the biblical to the bizarre happened so fast that you could miss the segue and figure his tale must be in the Genesis story or the phantasmagoria that was The Book of Revelation.

We often stopped at Uncle Albert and Aunt Marie's after church on Sunday night on our way home. We were already dreaming of wondrous things, our imaginations charged by powerful biblical stories. Indeed, among my fondest childhood memories is sitting on Sunday evenings beside the open windows in our little white frame country church located in the corner of a pasture where I could listen to night sounds as evening came on: cicadas, lowing cattle along the fence, the whoosh of the wings of swooping nighthawks, all alighting in my evening reverie. Inside the sanctuary, these sounds were interspersed with the familiar lyric four-part harmonies of the voices of my community singing softly on Sunday nights the hymns we all knew by heart. As I look back from this distance, I recognize that I would never again know moments so mellow and peaceful.

Sitting at Aunt Marie's table alongside the large swinging pendulum of the deep green, flower-painted face of the Kroger wall clock someone in the family had brought from Russia, we held on to the Sabbath as long as we could before we faced the work of the week. Sometimes we checked out baby geese or ducklings or the goats they had. Sometimes we gathered around the old piano and sang gospel songs from an old green book with shape notes. Sometimes we just ate and talked of the church news of the day.

On just such a Sunday night we got the deep sea creatures story after Uncle Albert had returned from visiting his Colorado brother, including their trip to a Denver natural history museum. With his typical passion he described the reconstructed skeleton of the dinosaur he had seen at the museum, elbows leaning into the distressed wood of the dining room table, and recounted his own vision of when dinosaurs and all manner of flying beasts and lizards had inhabited this earth. He compared their size to our barns. One step of such a prehistoric beast would have flattened his pickup truck, he said. In the end times we should expect these creatures to return again.

It must have been getting late and we were probably all tired as we sat around the table eating Sunday dinner's leftover cold pork and the *Schnetje* (a scone-like pastry) Aunt Marie made with lard. After a long day of church we were prone to think otherworldly thoughts anyway—certainly now, as Uncle Albert's eyes got wilder than any preacher's I had ever seen. His authoritative bass voice softened, lowered and then soared dramatically, describing the distant past as if he himself had run from those fast-footed beasts.

A little pot-bellied by then, he hunched over the table, waving his weather-scarred, work-abused hands. From a family of cowboys, all the men in his family kept that leathery, wise, Will Rogers look. Uncle Albert's tiny white mustache quivered over his small mouth, but I was mesmerized by his watery blue eyes. I believed those eyes really had seen the malevolent creatures of the deep, wrinkled with horn-like scales and tiny eyes and gnashing teeth. And he, maybe all of us if it was soon, would see them again when Christ returned in the Rapture and the world as we knew it would disappear—burn or flood or give itself again to sea monsters.

On Innocence:

Now jaded by experience, I think often of my childhood innocence. Is that innocence I assumed related to farm life? Is it because of the way I see myself as a Mennonite? I am tempted to paint my childhood as an idyllic life of innocence on the land. Do I preserve that sense of innocence so that I can see myself as "the quiet in the land" and thus, not responsible for the sins of the world?

Is it endemic to our long tradition of living apart in isolated communities, protected, often separated from local politics? Is that our way to protect ourselves from our own complicity, for example, in our use of land that was once homeland for people driven away before we arrived? Certainly, my parents did their best to preserve my childhood innocence, keeping me from worry and fretfulness. They did not share their financial struggles. I was unaware of their strains with family or community members. I knew nothing of the debt that weighed heavily after the crops that did not make. They protected me from the traumas they must have absorbed.

In my 1950s childhood, my job was to do well in school, to become educated. I didn't work on the farm much during the school year. My job was to fulfill my parents' dreams, to become what they had longed to be. College-educated. Teachers. They worked as hard as they did so we could move on to something else. So that we could move up? My parents found a way to give us what we wanted. Really, we lacked for nothing. Were they filling our empty stomachs to assuage the hungers of their Depression childhoods?

Despite the condemnatory scorchings by all the preachers I ever heard during my childhood, I never felt like the sinner they told me I was. Oh, I see my sins. I can count them. I beg for forgiveness. But I do not recognize myself as fallen. I attribute that to my sense of self as one with the natural world, and that natural state is as it should be. Is that related to innocence, a refusal to feel guilt?

Early 1950s

Chapter Four

Sand

It is summer, a hot, fly buzzing, lazy afternoon. No one has asked me to do anything. My little sister is taking a nap. The dog—the collie named Dusty—is stretched out full length on her side in the hole she has dug for herself in the cool sand in the corner against the concrete step of the screened back porch. Dusty, memorialized in our family lore for saving my little sister when she wandered off at play near the edge of the deep sewer hole behind the dairy barn, fiercely forbidden territory for us as children. According to her friend, as my three-year-old sister began to slide down the loose sandy precipice into the hole, Dusty grabbed at her clothing with her teeth, and like any good Lassie, dragged her back to safe ground.

On this particular hot summer afternoon, my bare toes are dug deep into the cool sand under the large Chinese elms that shaded our back porch. Grass does not grow here under the trees; instead of mowing, we clean and rake the sand into even furrows before company comes. The sand is both canvas and playground; it sculpts and molds beautifully. My sister and I construct long elevated highways with packed sand for our toy cars and trucks. We make angel food cakes out of moistened sand in our playhouse. We dampen and shape the sand into shapes we display on our weathered barn board counters, flipping deftly but carefully the molds we use to make stars and moons and chickens.

When a hard rain pounds the sand, we leave deep footprints in our own beach alongside the puddle in front of Daddy's tin shed where we float flat wood boats. Sand is the pit into which we dive as we practice our high jump every spring to prepare for the track meet. Sand defines our play and our family's existence. The windmill makes its familiar, sporadic metallic clicks as it adjusts its fans in the breeze. Barefoot in the sand, I am in that moment unworried about three-horned stickers or sand burrs, the two most dreaded dangers of walking barefoot on our farm.

Geologists have described the folding and faulting that occurred in southwestern Kansas in the early Quaternary period before, in Pleistocene time, streams spread fine sand and silt all over the Cimarron River Basin area. Sand, "loose, unconsolidated rocks having particle sizes between those of silt and pebbles (1/16 to 2 mm)"[4] spread throughout Kansas. Seward County has been described as "loess and river valley deposits" and "sand dunes."

Decades after we left the farm, I would discover in my parents' files a soil conservation plan and agreement dated August 21, 1952, constructed for the 960 acres my father farmed initially as "operator" for the owner of the land, L. G. Blackmer. "The soils consist of sands and sandy loams on a nearly level to hummocky topography having undergone moderate to severe wind erosion in the past. These soils need a good crop rotation including grasses and legumes since they are very low in organic matter content."

The soil conservation plan includes directions for stubble mulching and suggests that the land not be burned or overgrazed, but rather worked with a chisel to maintain cover against blowing—undoubtedly, a response to the over-plowing that had produced the Dust Bowl. East-west strip-cropping was advised to prevent wind erosion and crops were to be rotated: milo, rye, and vetch with sweet clover following milo on sandier ground. The attached plan for planting trees included sixty each of Austrian pine, green ash, hackberry or cottonwood, Russian olive, and 175 multiflora roses.

The multi-colored land capability map made by the U.S. Department of Agriculture Soil Conservation Service designated in yellow the land suitable for cultivation with easily applied conservation practices; land colored red was suitable for cultivation with complex or intensively applied conservation practices; marked in blue was the land which should be in grass but which could be cultivated to a limited extent with complex or intensively applied conservation practices; land not suitable for cultivation but suitable for grass or trees was colored orange. Our land was red and blue, "low in organic matter."

Clearly, the plan was not binding; neither the owner nor my parents as operators had signed it. The trees suggested were never planted, to my knowledge. We inherited from earlier post-Dust Bowl years the Osage Orange hedge that bordered our farm's west edge to stand against the ravages of wind. Still, my parents were well aware of the difficulties they had undertaken with the sandy land they had agreed to restore to productivity.

Mama tells me today that in the early years the soil was simply too "sorry" to grow wheat. From my earliest recollection, the northernmost, least fertile and sandiest soil was planted either in Sudan grass for our cattle or ensilage for cutting and filling the silos; the center part of the half-section was always planted in milo, and the southernmost third of the land east of our house was eventually drilled into wheat.

In recent years there has been renewed interest in planting sorghum, that "camel of crops," an ancient grain believed to have been brought to the U.S. aboard slave ships from Africa. Sorghum has received new consideration by farmers the world over because of increasingly hot and dry conditions globally. Today, even farmers in much wetter northwest Kansas are planting the gluten-free, nutritious grain sorghum, which is becoming popular among consumers along with other ancient grains like quinoa, amaranth, and spelt. Milo, a grain sorghum my father always grew, tolerates poorer soil and needs one-third the water that corn demands, so it has always been more suitable for land like ours than wheat.

Daddy would not long continue farming the nearly 1,000 acres he had taken over with help from my mother's younger brother, who had come to stay with us after my sister's birth in 1953. Mama's version of why they downsized their farming operation is instructive about their values. Sometime after my sister was born, Mama tried to leave the new baby with Daddy while she went out; my sister cried all evening. Exasperated, Daddy said, "This is enough. My own child does not know me. I will give up farming the east section and maintain the half section at home with my dairy. I will not work so many hours that my own child does not see me enough to know her own father."

That section of land east of us that Daddy gave up was later irrigated, something my father would never have considered, for reasons of stewardship of the land, expense, and work. Instead, my parents built the forty-cow family dairy, and Mama and Daddy headed to the barn together twice a day for milking. As part of the family operation we girls were assigned to make whatever Mama had planned for supper while our parents did the evening milking. During the summer or in busy times we drove tractor as needed and assisted with the dairy.

Underlying the sandy soil of western Kansas is the Ogallala Aquifer, with springs which have served both game and people for 12,000 years. The Ogallala (a word that translates from its Sioux origin, "to scatter one's own") is one of the world's largest aquifers, underlying eight states or eighty percent of the High Plains, and supplying drinking water to 80 per cent of the people who live there. Formed two to six million years ago, the reservoir varies in depth from a few feet to more than a thousand feet in various areas in the northern Plains and recharges at a slow rate. A dry land farmer like my father spent his life trying to understand the production capacity of the sandy soil he farmed; he had to adapt his techniques to the land—a lesson still being learned today in southwest Kansas for those farmers still trying to farm the land who recognize now that irrigating will suck dry the water beneath them.

Perhaps rationalizing about the soil he had been able to find to farm, perhaps in real gratitude for the life that it gave him, Daddy always touted the sandy land's attributes as he disparaged other soils that clumped like clay. He noted that implements became mired in claylike soil which clogged and clotted. But of course, clay didn't blow. The blowing sand must have haunted my father from the day he first set foot on our plot of ground.

Like my father, I, too began to love the sand. It is clean—in essence, tiny particles of glass. It sifts beautifully through screen or sieve and pours like time in a bottle. Rain washes it into beautiful patterns. Snakes leave on sand a trail of perfectly legible markings. The water we pumped from deep beneath it was sweet and cold. We loved the taste of our water, not like that nasty "gyp water" my parents had drunk from below the red dirt of central Oklahoma where they grew up. And a farmer who knew what he was doing and caught some luck could raise crops in the sand. With practice, it furrowed straight. My parents knew they only got their chance at farming this particular half section we would call "home" because of its history of blowing away.

Our landlord Doc Blackmer, banker, horseman, polo player and medical doctor, had come to Hooker, Oklahoma (named, I have read, for John "Hooker" Threlkeld, a Kentucky roper known to the Oklahoma Panhandle as so good with a rope he could drop a loop over a calf beside its mother and "hook 'er" in a heartbeat), in 1914, founded the First National Bank, and remained its president until 1952. It must have been in his role at the bank that Doc Blackmer bought up, foreclosed on, or somehow gained title to land around Liberal forsaken by owners during the Dirty Thirties.

Doc Blackmer was my parents' hero. He charged my parents no interest; they paid him when their crops came in. He opened a line of credit in the Liberal hardware and lumber stores so that they could make the place habitable. Almost immediately after they moved across the border into Kansas, they started the dairy that sustained them, providing the monthly paycheck when the crops failed. They cleared the twisted trees that covered the place, took down the ramshackle buildings, and used the wood to build calf sheds and barns and chicken houses. It is said that there were no ghost towns in the southwest part of the state of Kansas because every stick of wood on a landscape very nearly treeless is so precious that when a town died, every piece of it was reused, carried away, and recycled.

Sometimes Doc Blackmer kept a horse on our farm. His were magnificent horses, not like the old plugs Daddy kept on the farm for us girls to ride. Mama still remembers how she worried that Daddy would be thrown by a handsome black stallion Doc Blackmer boarded on our farm and wanted Daddy to help him break. Years later, I would read in the newspapers how Doc Blackmer once paid $10,500 for the legendary Appaloosa named High Hand, a racing, cutting, and roping horse that eventually sold for more than six times that price.

Doc Blackmer organized a polo team and had the first closed car in Hooker. In an interview he said that in his lifetime he had probably seen 300,000 patients from as far west as Springfield, Colorado, as far east as Beaver, Oklahoma, as far south as Texhoma, and as far north as Plains, Kansas. He died in 1960 before he had reached the age of eighty, but I remember him as an old man during the 1950s when he came to see Daddy, dressed a bit like a dandy, I thought, in a red plaid sports jacket over cowboy pants and good cowboy boots, his gray hair curling down his neck beneath his polo cap.

When he drove onto the yard, Mama hurried to call Daddy from wherever he was. I watched the two men stand together at the corral watching the horses, Doc Blackmer with his arms over the top rail and one leg up on the bottom rail, Daddy erect in his striped overalls, white T-shirt, and work shoes. There they laughed and talked together, this banker who had taken a chance on a couple of twenty-four-year-olds eager to build a life on a half section of land that looked to be beyond reclamation, and my tall, lanky father with a military demeanor: erect, his shoulders thrown back with pride.

A month after I was born in January 1949 and only months before my parents found the land, the Soil Conservation District had been formed, covering all of Seward County's 408,960 acres of land. The county signed on with the federal government for assistance with two projects: irrigation systems and re-vegetation of depleted rangeland and cropland unsuited for cultivation. In February 1954 terrible winds returned conditions in Seward County temporarily to those horror days the settlers had experienced during the Dust Bowl years, but most farmers in 1954 immediately began emergency tillage operations, and wind erosion was controlled.

My father was conscientious, a careful student of the sandy soil he farmed during the quarter century he farmed it, and every farm journal or periodical he received was filled with soil conservation ideas, many of which he employed: planting Sudan grass, summer fallowing, using strip crops, chiseling the soil. In 1958 Kansans harvested the second largest wheat crop in the state's history, bringing in cash receipts of over one billion dollars to the state's farmers and ranchers. Mama and Daddy paid off the debt on their sandy farmland with their earnings.

Part II

The Blessing of the Seeds: Midpoint between Winter Solstice and Spring Equinox

In Mexico, we celebrate with the locals in Queretaro the festival of Candelaria, the old tribal ceremony for the blessing of the seeds in preparation for spring planting. They share their tamales (from the Nahuatl for "wrapped food" and dating back to the Aztec maize offering to the god of water to ensure rain for a plentiful harvest).

In the U.S., Groundhog Day has become symbolic of turning from winter to look forward to spring.

In Liberal, we have Pancake Day on Shrove Tuesday. All these celebrations grow out of a basic human need to anticipate spring as people tire of winter and begin preparing the earth for spring planting, even in their urban back yards. How long will winter last? It is a seasonal concern that directly affects the farmer or dairyman.

In Kansas whether one is aware of Lenten practices or not, we need Shrove Tuesday, for cabin fever has set in after three months of winter. Seed catalogs arrive; we plan a garden; we eat pancakes and listen to the state news to learn whether Liberal has won the Pancake Race over Olney, England this year. We long to establish some Lenten discipline—if not to give something up, at least to perform some ritual turning to follow the lengthening day: six weeks until Easter when we expect to be somehow resurrected with the earth.

1953

Chapter Five

Wind

I think of the wind as the Navajos do, as Holy Spirit. In *The Spell of the Sensuous*, David Abram's argument for human inhabitation of the natural world to regain our senses, he reviews the Navajo belief that all air is Holy Wind, human breath included: "Wind existed and provided both breath and guidance to the other Holy Ones, such as First Man, First Woman, Talking God and Calling God."[5] For the Navajo tribe, wind is responsible for both creation and conception. "That which is within and that which surrounds one is all the same and it is holy."[6]

The biblical account of the Holy Spirit wind in the Pentecost story when Jesus' disciples huddled in an upper room after his death records "a sound like the rush of a violent wind." It filled the house and those gathered there; the energy and comfort of the Holy Spirit came to them as wind. Of the Trinity, the Holy Spirit wind was and is, for me, most real: the deep sensory and visceral understanding of the deeply spiritual all around and also within me.

For the Navajos all the winds of earth are palpable: "the actual gusts, breezes, whirlwinds, eddies, storm fronts, crosscurrents, gales, whiffs, blasts, and breaths" are "momentary articulations within the vast and fathomless body of Air itself."[7] Even as a child I knew to unbraid the Holy Spirit wind from the Trinity—"the Father, Son and—" for my own female divine. To this day I love the wind on my face, my eyes tearing, my hair pulled away into air, my body's weight overpowered as I struggle against a force so mighty that it makes me remember who I am: a mere mortal.

As a five-year-old child during the 1953 and 1954 dust storms east of Liberal, I witnessed my mother's fear when a fierce wind turned day to night at noon. Though she had not grown up in the heart of the Dust Bowl where she later raised her children, she was close enough to have known to fear death by dust pneumonia, that scourge of babies during the Dust Bowl years, when small lungs filled with dust and breathing became impossible. I watched as she hung wet towels over our dark windows, glancing at my baby sister in her crib.

Later, my father's dark face in such winds also showed his respect for the wind's power to blow away the sandy land he was powerless to hold as he watched it begin to drift like snow under his neighbor's barbed wire fence. Did he fear in his heart of hearts what those who left this land before him must have concluded—that this sandy land was actually unsuitable for farming? Surely, he must have agreed on those days when the sky grew dark and he waited inside the house for the wind to settle, those days when I heard him say to my mother, "Close the curtains; I can't stand to watch." On those days he settled in the armchair and took up the latest issue of *Successful Farming*, the magazine he often used to distract himself; on such a day, he must have felt himself a transgressor on the land.

In 1953 and 1954 in the Dust Bowl territory including Liberal, when my father was still learning to farm the sand, conditions in southwest Kansas were not that different from the Dust Bowl days of the 1930s. Fortunately, much had been accomplished during the 1940s to keep the land from blowing: shelter belts and wind breaks had been planted; over-plowing had been corrected; land had been set aside from cultivation, held in place by grass or other cover crops. The rains eventually came, and disaster was averted. The best wheat crop we ever had—the one my parents must have taken as an endorsement of their efforts—was only five years later in 1958 when our land yielded 60 and 70 bushels per acre. Pictures in the family albums show my sister and me standing in that "bumper crop," only our heads and shoulders visible above the tall, magnificently thick wheat crop we would harvest.

A map of the Dust Bowl territory covering parts of six states is shaped like a left footprint. At the southernmost edge, the heel, is Lubbock, Texas. On the right edge is Shattuck, Oklahoma; the left side boundary includes Clayton, New Mexico, and Lamar, Colorado. The northernmost edge of the Dust Bowl extends as far as the Republican River and Willa Cather's Red Cloud, Nebraska. The heart of this area includes towns in my homeland area, Boise City and Guymon in Oklahoma, and Liberal in Kansas. As Timothy Egan says in his classic *Worst Hard Time*, "The families in the heart of the black blizzards were . . . in towns like Guymon and Boise City in Oklahoma, or Dalhart and Follett in Texas, or Rolla and Kismet in Kansas."[8] The small town of Kismet is my high school alma mater.

On Black Sunday, April 14, 1935, referred to as the worst of all Dust Bowl days, twice as much dirt blew away as was dug out of the earth during seven years of digging the Panama Canal, according to Egan: "More than 300,000 tons of Great Plains topsoil was airborne that day."[9] Two thirds of the population who had been in the area in 1930 remained—even through the worst of it—perhaps defining the character of those longtime farmers who remain today in Seward County in Kansas and Beaver County just across the border in Oklahoma.

Some 200,000 people migrated out of the Dust Bowl area around Liberal during the 1930s. The McDonalds who had homesteaded our land were gone even before 1930. Egan's description anticipates my parents' coming in 1950:

> Afterward, some farmers got religion: they treated the land with greater respect, forming soil conservation districts, restoring some of the grass, and vowing never to repeat the mistakes that led to the collapse of the natural world around them and the death of children breathing its air. Many of the promises lasted barely a generation, and by the time the global farm commodities era was at hand, the Dust Bowl was a distant war, forgotten in a new rush to spin gold from straw.[10]

Though my parents carried the optimism of the times when they came to the sandy land east of Liberal, they hardly had delusions of spinning gold from wheat straw. They simply wanted to "make a living" on a half section supplemented by the dairy.

The Dust Bowl had resulted from good times and over-extension. Liberal had been the broomcorn capital of the world before the vacuum cleaner ruined the market for broomcorn. Egan argues, "Prohibition saved the broomcorn farmers, making grain more valuable as alcohol than the dried stalks had ever been for sweeping."[11] My mother remembers that soon after my parents' arrival at our farmstead in 1950, as they began clearing the debris west of the house,the scrub brush grown up through years of neglect, they uncovered the remains of the old still which must have been part of that valuable alcohol production. I smile to think of my mom and dad, a fresh-faced and eager tee-totaling couple of twenty-four-year-old Mennonites, quickly disassembling the operation they would have seen as desecrating their own backyard.

My parents were born in central Oklahoma in 1926 during the height of the boom years in the Oklahoma Panhandle and southwest Kansas. During these wet years farmers were breaking up grassland like never before. Wheat prices were high, and money was easy, especially for the suitcase farmers who came, tore up the land, sowed it into wheat, and went home to wait for the harvest to show them the money, hand over fist. The lore has it that one Ida Watkins, "Wheat Queen of Haskell County" (just north of Liberal near Garden City) claimed she had made a profit of $75,000 in 1926—a sum which would have been more than any baseball player's salary except Babe Ruth and more than the president of the United States made that year.[12] The potential of the sandy southwest Kansas land was believed to be limitless.

However, by 1930 the McDonald family who had homesteaded our land east of Liberal had deeded it to Lucie J. Furstenberg; in 1934 it went to The Federal Land Bank of Wichita. There is a foreclosure on the Furstenbergs in 1942, and in 1944, Lucie J. Furstenberg and John B. Furstenberg lost the land to Doctor L. G. Blackmer.

And that is how it came to be that I would grow up in the 1950s with the Dust Bowl's Holy Spirit wind in my ears. My sister and I had a girls' nest, a bikini-bottom-shaped triangular area between the two large silos where we played, always to the music of whistling wind song as it rushed through the small opening where we wedged our way inside, moving sideways. On ferociously windy days I sometimes sought out the two largest tumbleweeds I could pull from a fencerow, maybe four feet in diameter, grasped the tough stems firmly in each hand, hoisted them high above my head into the wind to "fly" down the gravel road, my feet hardly touching the ground as I sang in full throat above the mighty whoosh of the wind. I can't remember when I didn't know Bob Nolan's song, "Tumbling Tumbleweeds," made famous by the Sons of the Pioneers.

Most sources believe that the tumbleweed was inadvertently imported into South Dakota from Russia in the 1870s—brought in with flax seed by Ukrainian farmers like my ancestors. This Russian thistle takes root in disturbed soil like that opened by the railroad as it crossed the plains. The tumbleweed sucks water out of the soil at an outrageous rate: one Russian thistle competing for water in a wheat field is capable of removing forty-four gallons of water in a year. Its size can vary from a soccer ball to the size of a Volkswagen Beetle, this "wind witch" galloping across the flat ground as it disperses its 250,000 seeds. A typical tumbleweed around Liberal while I was growing up was three feet in diameter.

The quintessential wind image of my childhood was the sudden quickening of a whirlwind appearing magically as a kind of supernatural visitation. Sometimes called a "dust devil," it seemed to personify the wind's mysterious nature. It stirred and scuttled, roiling the sand and growing taller as it swept over the flat plain. As a young child I was both frightened and mesmerized, horrified and fascinated. I felt my body tense up in wonder at this mysterious and exhilarating stirring of the wind.

Indeed, this sudden phenomenon could send a shiver up my spine when it happened on a still day. Years later I would read of the *duende*, described by Garcia Lorca in his classic essay as the very base of creative energy come up from the earth: "Through the empty archway a wind of the spirit enters, blowing insistently over the heads of the dead, in search of new landscapes and unknown accents: a wind with the odour of a child's saliva, crushed grass, and medusa's veil, announcing the endless baptism of freshly created things."[13] Lorca describes the wind as arriving out of the ancient culture "of immediate creation" as if it were the spirit of the earth. He links *duende* with human longing as it "draws close to places where forms fuse in a yearning beyond visible expression."[14] For me the mystery of the wind, especially the whirled wind traveling over the sand, represented my own yearning for the Spirit come up out of the center of the earth.

Today, if I were to visit our farmstead east of Liberal as a trespasser, I imagine that what would be unchanged would be the wind between the silos. Always, we were at the mercy of the wind which sometimes behaved like an angry god. I see my mother's face—dark as the skies while she grimly hung wet towels over our windows to keep the blowing dust out of our house. I see my father's dark look as he watched his blowing fields, knowing he would need to disk up crops to try to save the land. Or, when he came in after furrowing in his attempt to hold a patch of blowing dust, his face black with dust. My father's sudden death at the age of fifty-two of a rare pulmonary disease of unidentified origin caused us to ask ourselves—could it have been the dust he breathed that took his breath and stopped his heart?

1957

Chapter Six

Rain

Scientists have coined a word for the scent of rain on dry earth, the wet smell of dust after rain: "petrichor" from the Greek word for stone, *petros,* plus *ichor,* the fluid ("blood") that flows through the veins of the gods in Greek mythology. I know that scent, the lifeblood from the gods scientists say is actually the smell of oil that plants exude in a dry period absorbed by soils and rocks and released into the air when it rains. For me, it is the smell of wet sagebrush. It doesn't often rain east of Liberal. We lived our lives wishing and praying for rain. When the rain came, we were in spiritual ecstasy.

Describing a desert rainfall in Chaco Canyon, Craig Childs in *House of Rain* says, "It was incense to me, a lurid scent that I have encountered only select times in my life, brief hours of the desert erupting into sudden and monstrous floods, where everything living and dead is channeled into a single slot. It smelled like creation itself."[15] In an outrageous act of exhilaration (and danger) Childs opens his book with a description of throwing his body into the flowing desert stream to be carried by the stream for miles.

I understand the sensory thrills Childs registers from my own "desert" childhood experiences. A rain's rare smell and the exuberance even a small shower elicited for us help me to recognize that I grew up on semi-desert land. Water was precious, sweet to the taste, to be treasured, almost revered. It connected to the divine. My daddy would say as he longed for rain, "Let's do a rain dance. Maybe God will send rain." Our ritual communal prayers in church seemed always to include a petition for rain.

Even the prospect of rain made for extraordinary behavior at our house. If a thunderhead arose in the south or west and seemed likely to come our way, we watched with obsessive contemplation, placing bets like gamblers on its likelihood of hitting us, and exactly when, or fearing aloud that it would go south and miss us, showering instead the Oklahoma Panhandle where others in our church community farmed. Or maybe it would go north of Liberal where we usually speculated that they had had enough rain already. Typically, a storm cloud arose in the southwest and moved diagonally, toward the northeast. I have sorely missed this cloud-watching for rain since leaving the farm. I have never again lived on so flat a horizon, under such an immense sky.

We girls tried to learn the variety of clouds and their portents. What made a green colored cloud suggest hail to our parents? Why was that dark cloud sure to bring wind? How did Daddy know that another cloud's color and shape would never bring rain? Only rarely do I remember seeing a tornado-shaped funnel in the far distance. On the flatlands of southwest Kansas you can see in any direction for miles; you have time to make a decision about whether you are in danger.

One of my most precious memories of life on the farm east of Liberal is stepping out of the house onto the screened porch and then out the east door, whether by day or night, to check the sky for the telltale signs of weather, the sun's shadow, the wind's direction, or most probably the red, pink, or orange tones of a rising or setting sun unimpeded by clouds.

If it did rain, our bodies responded involuntarily, convulsing in joy. We danced or ran or jumped; we laughed and shouted. As the wind began to rush around as if trying to find a direction, we ran to close windows left open on the pickup truck, raced to shut a door in the barn's feed room, to cover a bin—excuses to feel the first stinging pelts of cold rain against our parched skin. Most exhilarating of all, the smell of the first drops of rain mixed with soil and grass, the wind bringing in the fresh scent of sage from the pastures, even the familiar musty smell of dry manure in the sandy cattle lot moistened by the rain. I have danced for joy barefoot in the rain. Even today, with an adult understanding of the danger of lightning strikes, I do not fear a thunder cloud pelting me with stinging drops of rain.

When we had the good sense to get inside the house before the sky cut loose, our family of four crowded together, peering out of whichever window offered the best visibility: "Look at how the sand is washing there off the cow lot!" Rain patterns on the earth were a phenomenon for which we had multiple descriptions. Our maternal Grandfather Kliewer had used words for different kinds of rain which my parents laughingly remembered in Plaut Dietsch when the rain actually poured down. Those words come to me today only when I observe the rainfall phenomenon they describe. If the rain ridged the sand in a downpour, our parents told us to look at how the water had "shaled" to create the washboard effect. A word which always tickled my parents and apparently referred to rain bubbling into earth was one they also attributed to Grandpa Kliewer: "*Kjikje 'mol doa, de pluusta*"—meaning something like, "Look there how the rain is bubbling and gurgling into the earth." A long rain that lasted more than a day was termed a "Suada Raajen" (soo-da ray-en), the rare, slow, gentle rain which soaked the earth.

To this day rain fills me with hope. I sometimes think it would be a good way to die—in a thunderstorm, struck by the lightning of such promise! Daddy's mood, which was eternally upbeat despite the odds he faced on the farm east of Liberal, turned to boisterous joy if it rained, and he might kick up his heels as if he were one of our baby calves high tailing it. He might wave his arms in a farmer's Tevya dance in his pin-striped overalls, whoop and jump to try to click the heels of his work shoes. And when the rain came to southwest Kansas and the Oklahoma Panhandle, we were profuse in our thanks to God during our Sunday morning communal prayers: "for the rain which in Thy benevolence, Dear God, Thou hast bestowed upon us in our need." We must have believed that if we were grateful enough, there would be more.

On our farm, however, it was a lucky day if the rain lasted long enough for my father to say, "I'll bet we got an inch" before he headed for the rain gauge. Daddy always felt that he needed an inch of rain to make a difference in an area that averages less than 20 inches per year. During the worst years in Seward County, annual rainfall totals could hover around 10 inches. Childs puzzles in *House of Rain* over what sort of exuberant celebration among the Chaco Canyon people might have led to "eighty thousand ceramic vessels intentionally smashed at one site. . . . like casting champagne glasses into a fireplace, an exploit of stunning abundance and celebration in this crumbling, bare-bones land."[16] I knew instinctively that it must have rained!

Daddy always took the occasion of rainfall to bless the sand he was grateful to farm. Our newly planted seeds did not wash out with a rain as easily as other farmers' crops planted in muddier soil where the seeds might become mired and buried, or the soil might wash over the seed to form an impenetrable caked crust. Sand did not "wash" or move as easily. And the cows' lot was sandy, which made a good base for cow manure to dry and compost, almost as if it were being filtered, rather than to muck up as in other types of soil in a rain. Our dairy cows did not mire down in the sand nor track sticky mud into the barn and create a mess which we would need to hose down in our Grade A dairy barn. They carried in a devil's claw (from the unicorn plant) clamped around a hoof more often than they carried in mud! Daddy often noted that a dairy cows' pen on the sand was a blessing. Our cow lot mixed and sifted sand and manure so that Daddy could haul it in the manure spreader onto our fields and spread it over the crops as fertilizer.

For farmers and ranchers rain is almost always a welcome respite in life's perpetual rhythms, a relief, not only for the earth and its creatures, but also for the caretakers. If it is pouring, farmers must race home from the fields with freshly washed farm equipment and take the rest of the day off. In our case, though, a rain might give my father a rest from his tractor driving, but there was never a break for the dairyman or woman. Twice a day, morning and evening, rain or shine, the cows had to be milked.

Because of the scarcity of rain in our area, early plains inhabitants would not have thought of settling a spot like ours east of Liberal on the open plain to plant or grow crops; they would have situated themselves along the river where they had access to water. But settlers like us put down roots, drilled into the earth to find water, and waited on the rain to come and water our crops far from the river.

Driving Highway 54 over the Cimarron River's mostly dry bed each day enroute to school in Kismet on the bus, I had no idea that Coronado had camped alongside the river 400 years earlier. Coronado had named the Llano Estacado or "palisaded plains" frequently mistranslated as "staked plain," that area of eastern New Mexico and northwestern Texas which constitutes one of the largest mesas on the North American continent. The table elevates almost imperceptibly at ten feet per mile from 3,000 feet in the south to 5,000 feet at its northern height. The Canadian River is the Llano Estacado's northern border; the plain stretches 250 miles to the south and encompasses 150 miles east to west, covering 33 Texas counties and four New Mexico counties. Coronado described it as a "sea of grass." That land and that view have always felt like home to me.

The Cimarron Basin covers 6,800 square miles of the southwest corner of Kansas. Viewed on a map the river's path forms a mountain-shaped peak over southwest Kansas at Ulysses, and then descends to cross Seward County diagonally, cutting our county almost precisely in half before it leaves the state and crosses into Oklahoma. The flow of that river is a much more natural boundary than the state lines which cross the river. Perhaps that is why Liberal and the land east of Liberal where I grew up always seemed more naturally a part of Oklahoma despite being north of the state line.

My favorite part of our Highway 54 bus ride to Kismet after our nearby country school was consolidated was our descent down the long hill into the Cimarron River valley through the blue and green horse country pastureland of sagebrush and yucca. I closed my eyes and pretended to be horseback, my head nodding along on the slow, hour-long bus journey home at four in the afternoon. Over one hundred feet tall and 1,200 feet long, the "Samson of the Cimarron" bridge paralleled our route on Highway 54.

The long valley we descended to cross the Cimarron contained no raging river in my time; it was mostly sand bar. Many years earlier, however, the Cimarron River had at least once swelled to become a dangerous rushing river a mile wide. They say that the flood of 1914 expanded the Cimarron River to rival the wide Missouri. The local area had experienced a five-year drought after which heavy rains occurred in the headwaters of the Cimarron in New Mexico. And then the rains became widespread; Liberal got three inches, and the stockmen along the river bank began to move their stock and their families as the Cimarron flooded its banks. Tom Ward, the railroad section man, tried to retrieve telegraph and telephone wires, and was swept away. He was found days later three miles downstream. Other lives were lost attempting to cross the flooded Cimarron, some whose bodies were found more than fifty miles downstream. Liberal was marooned for four days, its bridges washed away.

A subsequent flood and train wreck in 1938 eventually resulted in the decision to build the Rock Island Railroad Bridge in 1939 at a cost of 1.5 million dollars. Purported to be the largest of its kind, on pylons drilled to a depth of 165 feet, it was built to resist the shifting quicksand of the Cimarron River.[17] We school kids whispered what we had heard as we drove by the bridge every day on Highway 54: there were dead men buried in those enormous expanses of concrete poured to hold up the bridge, men who had died as heroically as war veterans to build the massive pylons that held that bridge high in the sky.

The Noble Farmer

The sand. The wind. The rain. These three features textured my childhood on the farm. Even though I read today of how differently Native peoples viewed use of the land in contrast to settlers like us who farmed it, how they tried to live with what the land provided rather than converting the land to their own uses, I would argue that we had a great, maybe even similar, appreciation for the land. When I read of Native peoples' theology of dependence upon the earth, I feel a kinship to their recognition of the sacredness of Mother Earth. We did not feel like conquerors or subduers or rapists of the land. We were not expansionist. If we tended it well, what we had would be enough because of the dairy. When I drive by those billboards proclaiming, "A farmer feeds 137 people plus you," I still think of farming as a sacred profession.

We were taught that we were caretakers, mimicking God, who sees the little sparrow fall. Nurturers. Stewards. We knew we could only remain on the land if we learned how to work within the rhythms of the seasons, and only if the sand, the wind, the rain allowed us to remain. We had the utmost respect for the sand we cultivated, the wind we knew could overpower us, the essential rain we could neither predict nor command. Though we were "making a living," dairying and growing crops to feed our cattle or to market and buy groceries, our work did not feel like commodification of the land. Each of our forty dairy cows had a name and a personality. We felt our mutual dependence—the cows on us, and we on them. Our days were filled with the good sweat of manual labor and joy, the smell of the soil, sunrises and sunsets, a waxing and waning moon, plants springing up in their season, the births and feeding of new calves, the mellow breezes of evening, thunderstorms, meaningful work as a family unit to complete the day's chores. I know that farming has become a big corporate business; yet, it is hard for me to see the farmer as I knew her or him, as other than a steward of the land.

No doubt, we were haunted by the prospect of failure on the land; maybe that is why we came to think of the land's offerings as gifts. Our sandy farm helped us to understand the sacred nature of water. Drought was always a possibility. You must pray daily for rain; that fact encouraged mindfulness. Our well water, sweet from the earth, was plentiful enough to sustain a Grade

A dairy barn where we hosed down the concrete daily. We were in the habit of offering water blessings. We knew the disciplines of depending on the land and the deep humility of having little control; is it possible that the subsequent gratitude was sweeter too, arising from a deep place in the soul?

Some suggest that after 10,000 years humanity may be on the cusp of a paradigm shift or further evolution in the practice of farming. The Land Institute in central Kansas argues that since its beginning, farming has depended upon an annual monoculture, three annual grasses—wheat, rice, and corn—to sustain humanity. But, for example, in a state like Kansas, today farmers are at risk of losing their crops due to the climate's warming and weather extremes, the degradation and loss of the rich soil that once existed here after centuries of perennial and diverse grass growth.

For two decades the Land Institute has been developing a perennial wheatgrass they call Kernza, in their belief that humanity will need to move to a self-sustaining perennial polyculture consisting of diverse and long-lived plants. Though no one has ever domesticated a grain that lived beyond a year, the Land Institute has determined to do so. Today, they bake bread with Kernza ("kernel" and "Kanza"), an intermediate wheatgrass they have developed to grow from much larger seeds than its wild ancestor, to have very long roots, and the power to pull 6.5 tons of carbon dioxide out of the air over an acre of soil and store it in its long roots. Regenerative farming practices today in the plains states have dramatically altered tillage practices; farmers use non-till methods as well as diverse plantings to restore the soil. Perennial polycultures would further affect the future practices of farming.

700-1200 C.E.

Chapter Seven

Plains Village Farmers

Everyone then who hears these words of mine and acts on them will be like a wise man who built his house on rock. The rain fell, the floods came, and the winds blew and beat on that house, but it did not fall, because it had been founded on rock.
—Matthew 7:24-25

In the area along the Beaver River straight south of our farm in the Oklahoma Panhandle, farming had been going on for a couple of thousand years after it began to accompany foraging. The earliest farmers began to stay in the same place long enough to raise a crop of pigweed, goosefoot, squash, and/or sunflowers.[18] In this early farming period crops like sunflower, native squash or gourd, may grass, marsh elder, goose foot and pigweed were domesticated.[19] All cultivation was done by hand with a hoe. While these farmers' predecessors as foragers had used skin bags and baskets, light and easy to carry as they moved from place to place, permanent dwellings meant these village farmers could make and use pottery, probably shaped by the women.

In the 1970s the University of Oklahoma excavated the site once used by a group of Plains Village farmers along the Beaver River on the Roy Smith Ranch where our longtime church friends, George and Shirley Kroeker, have lived for decades. The archaeological site in Beaver County is fifteen miles southeast of our land. The Village or Plains-Village inhabitants (A.D. 700 to 1500) may have traded with the Chaco Canyon inhabitants to the southwest; they almost surely had contacts to their north. These farmers grew crops in the sand along the Beaver and Cimarron Rivers 600 to 1,000 years ago.

The Beaver County terrain known as the Breaks is rough broken land between level lowlands and uplands. Erosion has produced its jagged broken look, probably not much changed for hundreds of years. Bumping over the cattle guard onto the rocky lane leading to the ranch house, the never-broken-by-the-plow grass-covered hills, the Beaver River, and small springs on the north side of the excavated archaeological site wind us backward through time.

The same open horizon must have been familiar to those who settled on this creek and its springs hundreds of years earlier. They used the stone foundations to build housing complexes: rectangular sandstone slabs three feet high formed walls that rest on bedrock and are held upright by clay—packed firm at the bottoms of the stone slabs. The shared west wall of the excavated dwelling is seventy feet long and sectioned into fifteen-by-eleven-foot rooms, living quarters for perhaps five different families. Holes were pecked into the bedrock for support poles used to hold up the thatched roof.[20]

The excavation team found bone and flint pieces, probably used for chipping tools and working on hides in rooms the archaeologists suspected had been kitchen areas. The round rooms were used as granaries for storing crops; tools for farming found on the site included bone hoes and knives. The inhabitants here were both farmers and hunter gatherers, using bows and arrows and spears, and mostly living on bison. The Alibates flint they used for arrow points came from the Texas Panhandle 100 miles away. Their grinding basins were sandstone.

George pointed out several grinding stone sites still evident atop the bedrock beside the Beaver River, dough bowl indentations made by grinding done 800 years ago. New Mexican rough stones were also found here. Excavators speculated that Plains Village Farmers like these in the Oklahoma Panhandle traded their bison hides, meat, Alibates flint, or crops for obsidian. Beads found here made from Pacific Coast shells suggest trading chains extending all the way to the Pacific Ocean.

The use of stone in home building suggests influence by the Pueblo Indians of New Mexico in places like Mesa Verde, Chaco Canyon, and Pecos. A potentially related group, the Upper Republicans in north-central Kansas, used similar tools.[21] They made grey cord-marked pottery with a round base characteristic of both the Upper Republican people and this group, whereas Pueblo dwellers in the Southwest made more decorative pots.

Their cord-marking was done before the clay was completely dry, using a smooth rock in one hand and a cord-wrapped paddle in the other, and beating the exterior with the paddle. A woman's hand inside the pot kept the clay wall from collapsing. Cord-marking may have thinned the walls to make the pot lighter and easier to carry, and sealed the coils of clay the potter had laid one atop the other in the first shaping of the pot. It is also possible, of course, that like women throughout history, these early potters simply liked the look and feel of a cord-marked pot—their cord-marking technique purely aesthetic.[22]

We crossed the creek on the Roy Smith Ranch on George's four-wheeler and crawled through brush to find the place where the tin cup had always hung on a peg in the rock for the thirsty cowboys going along this trail many years after those first Plains Village Farmers lived here. Of course, we would have been better served on horseback. My California-reared husband remembered aloud his first Oklahoma horseback experience on this ranch in a cattle roundup. You can still see the swales on the trail north of the creek, old inscriptions on stones, cowboys leaving their mark along the Adobe Trail. The spring no longer bubbles; the creek itself is almost dry.

This is quintessential No Man's Land, that oblong strip 166 miles long, east to west, and 34 miles wide, north to south—unclaimed after Texas (1845), New Mexico (1850), Kansas (1854), and Colorado (1861) had each marked their boundary lines on the Plains. This is unfarmed ranch land not far from our church. Plains Indians hunted here before the formation of reservations, but by 1870 cattlemen from New Mexico, Texas, and Kansas ranged over it. Two major "highways" were worn into it by hundreds of thousands of cattle driven from Texas to Dodge City over the Tascosa-Dodge and Jones-Plummer Trails.

In 1886 President Cleveland opened No Man's Land for squatters, and migrations into the area produced a temporary population of 15,000 people, most of whom departed when the lands to the east were opened. When Oklahoma Territory was established in 1890 and the Rock Island Railroad was built from Liberal to Dalhart, Texas, the old No Man's Land, now become the Oklahoma Panhandle, grew to a population of over 35,000 by 1907.

The Roy Smith Ranch was earlier the range of J. J. Fulkerson & Brother on Sharps Creek at the Adobe Walls Crossing. Charles Edward Hancock recalls, in his autobiography *The Call of the High Plains*, a week he spent as a child in summer 1886 or 1887 picnicking and swimming in the pool on the creek at the Roy Smith ranch site. Hancock remembers that they dug lots of cottonwood tree sprouts to take to their homestead in Stevens County in Kansas:

> So vast was the prairie as we made that journey across that flat land that we marked our trail with sod cairns, cut out of the prairie with a spade and piled up so we could find our way back home. That was when I first learned from the older couple who chaperoned us why the Staked Plains were so-called, and how the Spaniards under Coronado used stakes, poles, buffalo skull cairns, and other means to mark their passage across the treeless table land, so vast and flat that no man can tell where he is from one day until the next without having some guiding marks to follow.[23]

Oklahoma Panhandle history shows how closely the town of Liberal is tied to the area to its south, how rooted Liberal is in Oklahoma lives and migrations, and the arbitrary nature of drawing state lines. George and Tom Smith bought this ranch before the turn of the century. They opened a drug store in spring 1887 in a small (and short-lived) town seven miles southwest of what is today Liberal. In 1888 they moved their frame store building to Liberal, then a new town site being developed by the Kansas Town and Land Co.

Smith's Drug Store, one of the first and oldest continuing businesses in the county, was taken over by Tom's youngest son, Roy, a pharmacist, in 1934 and served as our family's drug store during my childhood years. Actually, it also served as our medical facility as we rarely saw a doctor or dentist. Daddy went to the pharmacist to ask for coke syrup to settle our stomachs if we threw up; Mama bought Vicks Rub for a bad cold, Mentholatum for congestion, liniment for an ache. Smith's Drug Store was sold the year I graduated from high school in 1966.

The Roy Smith Ranch had one of the first telephones in the area, line run on fence posts. Today the ranch house George and Shirley live in is still nestled down along Sharp's Creek, very near the old Adobe Trail which George flagged for us across the pasture where he could still detect it. The most obvious landmark still apparent is the corral which was part of the Bartholomew Crawford Road Ranch in the 1880s and apparently worked like a modern-day hotel.[24] There were probably both a wooden hotel and bar or restaurant here along the Adobe Walls Trail. Though researchers did find horse shoe nails, bullets, cartridges, broken bottles, and harness parts, the excavation produced mostly prehistoric artifacts which tell the story of a much earlier period.

While the Plains Village farmers lived along the Beaver River, the philosophical underpinnings to drive Indigenous peoples off the land in North and South America were being hatched on the European continent. Both Cortes and Coronado carried with them the understanding about who has rights to the land, no matter who is inhabiting or otherwise using it, that has recently come to be known as the Doctrine of Discovery.

Christian denominations including the Mennonites have begun to recognize how this doctrine, which grew out of European medieval Christianity, afforded them—consciously or not—the settler privileges which eventually became firmly entrenched in the U.S. legal system. The inception of this conquest philosophy goes back to the Crusades to the holy lands and the dispossession of "pagan" peoples. During the same years that the Plains Village Farmers were living along the Beaver River, Pope Innocent IV in 1245 determined that Christians could legally take invaded territories in defense of Christianity and the Church in a "just war."

Eventually, this framework would be developed and expanded, especially in the Papal Bull Dum Diversas in 1452, which established Christian dominion and subjugation of non-Christian "pagan" peoples. Indigenous peoples were to be subdued as "enemies of Christ" and reduced "to perpetual slavery." All of their goods were to be appropriated for the use and profit of the discoverer and his king. By the time that Cortes landed in what would become "New Spain" in the Gulf of Mexico in 1519, the rights of the "discoverer" were assumed.

Christendom's history implicates all of us who settled land in the U.S., all of us who took our colonizers' rights and privileges from "pagans" driven off the land. If we did not have Coronado's insatiable obsession for gold that drove him to pillage and slaughter, we Mennonites must still contend with our own prosperity ethic.

I have slowly come to recognize the implications of the long history of farming among my Mennonite ancestors. Wherever they have lived, they have converted the land to "productivity." They have altered its natural state—whether they drained the marshlands in Prussia, converted the grasslands the Nogai people used in Russia to farmland, or plowed up the grasslands in Oklahoma and Kansas. There are implications of altering the environment over the course of hundreds of years that no one generation recognizes, perhaps until it is too late.

Immigrant or Settler?

When I first read about the people known as Plain Village Farmers along the Beaver River in the Oklahoma Panhandle, I said to myself, "Ah, these are the first settlers in our area." They lived on and must have defended this site for generations. I have come to recognize my misunderstanding about the definition of a "settler."

I have long struggled with a central tension I have felt about my own people who came to this country to farm. Are we part of the immigrant tradition, those dispossessed and evicted, seeking a new homeland as did my Mennonite forbears, or were we empowered white settlers moving onto someone else's homeland? Was this simply an issue of origin or skin color or geography? The distinction is clarified by Eve Tuck and Wayne Yang in the recently published workbook written for settler accountability, Elaine Enns and Ched Myers, Healing Haunted Histories (HHH).[25] *"'Settlers are not immigrants. Immigrants are beholden to the Indigenous laws and epistemologies of the lands they migrate to. Settlers become the law, supplanting Indigenous laws and epistemologies'" (11).*

Enns and Myers contend that settlers are not guests who abide by the "law" of those hosting them on the land. Rather, settlers come to stay on the land. And settler descendants like my parents and generations before them continue "resettling" wherever they go. "Settling," then, is a mindset. This kind of resettlement is an ingrained characteristic of my Mennonite forbears for centuries, a pattern which eventually found my family in southwest Kansas long after the original peoples were moved away. I can hardly avoid complicity if I think for a moment how my Mennonite farmer forbears came to the land.

Sometimes it is hard for us as settlers to feel, acknowledge, or accept our complicity. I have often heard the argument that humans have been migrating since they began to walk. They are always and inevitably inhabiting someone else's homeland. Sometimes those who were there before had long ago moved away. Sometimes they have been pushed off. Are we then responsible for all of those who inhabited the land before us? If "settling" is a way of seeing the land and one's neighbors, we are. As Nikki Sanchez said in HHH, *speaking to resettled settlers like me, "[This history] is not your fault, but it is ab-*

solutely your responsibility" (quoted in HHH *12). I am coming to understand that as a settler, the paralysis of guilt over what happened on the land is not useful. What matters is my response to my recognition of who I am.*

Chapter Eight

Native Tribes

This is our inheritance, for those of us who imagine ourselves pioneers. We don't seem to have retained the frugality of the original pioneers, or their resourcefulness, but we have inherited a ring of wolves around a door covered only by a quilt. And we have inherited padlocks on our pantries. . . .
—Eula Biss, Notes from No Man's Land, *161*

I have tried to imagine the horizon which lay open to them as Coronado's party crossed the plains we would later live on in southwest Kansas. There were still 30 to 60 million bison in North America, animals the plains tribes considered kindred spirits, animals some tribes referred to as "the first people." The wolf was the bison's only natural predator in the wild but actually served mostly to cull the herd of the sick and the old.

The plains tribes had evolved to depend heavily on the migrating bison for their subsistence lifestyle, but they also built the rhythms of their lives around the bison's migrations. Never domesticated by the tribes, bison were grass grazers—their normal way to eat a couple of hours, rest, move on—in predictable migration patterns. It was a symbiotic relationship. The Indigenous peoples are credited with producing the lush grasslands which sustained the bison through their burning practices. "Native Americans burned the Great Plains and Midwest prairies so much and so often that they increased their extent; in all probability, a substantial portion of the giant grassland celebrated by cowboys was established and maintained by the people who arrived there first."[26]

European settler farmers, with their propensity for domestication of animals, tried, of course, to domesticate the bison. But their efforts were thwarted both because the settlers were unable to build fences to hold the bison, who could jump a six-foot fence, and because of the bison's speed: they could run 35-40 mph when roused and agitated.

In 2005 in Santa Fe, New Mexico, I stood above an exhibit designed to celebrate 400 years since Europeans began arriving for settlement in Canada and the U.S. The three-dimensional table map of the continent depicted three sites of nearly concurrent European arrival: Jamestown on the east coast, Quebec City farther north in Canada, and on the southern edge of what is today the United States in Santa Fe. The three cities vying for "first settlement" honors decided to provide a joint exhibit which showed that the three settlement groups had arrived all about the same time around 1605.

Looking down on the papier mache table map provided a strange vision, like floating in outer space to view the outlines of Canada and the U.S. I could see what has sometimes been described by Native Americans as the encroachment of the spiders (settlers) onto Turtle Island, their name for the continent. I could imagine, from this vantage point, the spiders clambering up onto the edges of Turtle Island simultaneously at the three sites depicted in Canada, the U.S., and what was once Mexico, rapidly increasing their numbers as their trails crisscrossed—east-west trails crossing north-south trails in the middle of the U.S. on the Great Plains. That unearthly view still haunts me.

The European farmer settlers would dramatically change the habitats where they lived, not only through plowing and farming which replaced the long practice of burning, after the much earlier introduction of the horse to greatly change tribal ways. Both human and livestock diseases were introduced, and guns would lead to the destruction of the bison. The natural interactions which had served both the tribes and the bison were disrupted by market forces, and Native American market hunters began to target bison, especially the cows who had better hides, for trade.

Several years ago I visited Fort Bent in Colorado, an important post during settlement times. A main feature of the tour was the buffalo hide press in the center of the fort grounds, where Cheyenne women readied the hides for shipment east by the ton. This cultural enslavement was in its time presented as a perk for the Cheyenne women, "honored" as the only members of the tribe allowed inside the fort.

Ironically, the railroads built in the 1860s laid their track on the worn trails of east-west migration, feeding routes the bison had created, even while hunters killed the bison they encountered to feed the railway crews and those quartered at the Army posts. On my first visit to the Coronado Museum on the east edge of Liberal, I copied into my notebook these words posted on the wall near the black and white pictures of mountains of bison bones: "The Union Pacific was completed into Abilene in 1867 and hides shipped east far more economically by train than by wagon. 'Buffalo Bill' Mathewson won fame by shooting as many as 1000 buffalo a week. During the peak winter of 1873, more than 750,000 buffalo hides were shipped out of Kansas, 400,000 from Dodge City alone."

In the year 1870 two million bison were killed on the southern plains. Bison bones were so cheap—$2.50 to $15.00 a ton—settlers burned them to refine sugar and make fertilizer. Hunting bison was great sport, leaving the plains littered with carcasses. Between 1868 and 1881 when my ancestors were arriving on the railroad in Kansas, 2.5 million dollars per year was brought into the state by the buffalo hide trade, the equivalent to the remains of 31 million bison. During each of the three years from 1872-1874, 5,000 bison were killed a day. It is no wonder then that by 1874, the year both my paternal and maternal great-grandparents got off the train in Kansas, the buffalo were gone.

It was obvious to the European settlers from the beginning that to settle the Great Plains they would need to rid the grasslands of both the native tribes and the bison they depended upon. The great Kiowa chief and orator Satanta also foresaw what farming would mean:

> We have always lived on [the land]. Then you came. First the traders. That is all right, for we were in need of blankets and kettles. Then the soldiers came. Then other men came. They are farmers. They want to work the land. That is not all right. The land doesn't want to be worked. Land gives you what you need if you are smart enough to take it. This is good land, but it is our land. You kill the land—when you do that the buffalo will never return. We have to save our country.[27]

The Kansas town of Satanta, 36 miles north of Liberal in Haskell County, named for the Kiowa chief, was located on cattle range until 1912, when it was opened to settlement six miles east of the Cimarron River by the Santa Fe Railroad. From open prairie to cattle range to railroad town to its claim to fame today—the county with the highest agriculture production in the state—the land of Haskell County has ironically proven the chief's claim that it is good land.

As the Arapahoes and Southern Cheyennes were being resettled on reserves and later on allotments in Oklahoma, the Mennonites settled on land among the Cheyenne and Arapaho allotments. Highly interested in education, the Mennonites recently come to Kansas would soon become involved in the Indian boarding school movement. The Quakers, who had the ear of President Ulysses S. Grant, argued for better treatment of the Indian tribes and called on their fellow pacifists, the Mennonites, to help set up a school for Cheyenne and Arapaho children at Darlington in central western Oklahoma. Bethel College in North Newton, Kansas, recruited its first president, C. H. Wedel, from his post at the Darlington school in Oklahoma.

After Darlington, the Mennonites were offered Cantonment, the former U.S. military barracks near Canton, Oklahoma, where they established Cantonment Mennonite Mission boarding school. We would like to think that the Mennonites' boarding schools were not built on the same "kill the Indian and save the man" education-as-assimilation philosophy as the boarding school prototype in Carlisle, Pennsylvania. We know that by 1900 thousands of Indian children, some as young as four years old, had been removed from their homes and cultural norms to attend 150 boarding schools across the U.S. where many young children died. If these children lived, the schools stripped them of their culture, language, family ways, and asked them to abandon their religion.

It is clear that the Mennonite schools were not free of the prevailing philosophy. In 1883 Samuel Haury, Mennonite missionary, reports to the Darlington Indian Agent:

> We may teach the Indian child all the arts of our civilized life, keeping him away from the influence of his ignorant, superstitious and idolatrous tribe for many years, but without a living Christ in the heart such a child, returning as a young man to his people, will soon fall back into the old superstitious customs and habits of his race. The Indians are a religious people; religion penetrates their daily life; almost every act that they do is connected with some religious meaning, scrupulously inculcated into the child from its infancy; and they will be civilized only by giving them a higher, the only true religion, that of Christ. (*Dismantling the Doctrine of Discovery Toolkit*, 27)

Of course, Mennonites' relationships with their Native neighbors were complicated. Missionaries were friends and allies too, preservers of culture and language. Mennonites were in a fight to maintain their own German language even as they settled the prairie states and played a role in boarding schools for Natives. The Swiss Mennonite missionary Rodolphe Petter, living in both Southern and Northern Cheyenne communities during his lifetime, was known to resist government eradication of Cheyenne autonomy even as he spent his life working to preserve the Cheyenne language in written form and created the first Cheyenne-English dictionary still used today. Petter is greatly admired by the Cheyenne people for his work preserving an oral language by transcribing it.

When my mother was a child in the 1920s and 1930s, her parents rented land allotted to the White Turtles near the small town of Corn, Oklahoma. From her stories I know that my German grandparents and the White Turtles from whom they rented the land had very different concepts of land ownership. When my grandmother Marie Kliewer saw that the White Turtles had come to set up summer camp along the Washita River on their allotment, part of which my grandparents rented to pasture their cows, she hurried to butcher chickens and put together food baskets for the White Turtles, who would soon appear at her door to request food.

Their expectation, I imagine, was that whatever had been grown on their allotted land should be shared. My grandfather was sometimes angry when he missed animals that he believed had been taken by the White Turtle clan. He clearly believed that whatever rent he paid the White Turtles was adequate compensation; he needed to keep his crops and cattle for his own profit. Whose was the produce of the land? Whose the river at whose stream the cattle drank? Who had been evicted? Who chose or knew how to farm? Who had the right to farm?

It occurs to me that in the view of my farming ancestors, if land was not being farmed, if permanent houses were not being built on it, if they saw no permanent settlement on the land, it was uninhabited. No one was there. They must have believed that they had a right to such land; apparently, no one wanted it. Such was the tunnel vision which sees farming as the only legitimate use of land.

My grandparents' grandparents had no doubt experienced similar disagreements in the Russian Ukraine where they had settled earlier, effectively evicting other tribal peoples. The Nogai people, a Turkish tribe, had long inhabited the Ukraine when the Mennonites arrived in the early 1800s at the invitation of Catherine the Great. Nomadic, the Nogai had been longtime suppliers of horses for the Russians, but were now forced to live a more settled life alongside the German Mennonites in the Molotschna region of southern Ukraine beginning in the early 1800s until 1860.[28]

Mennonite agricultural use of the land encouraged the Nogai to become settled sheep herders share-pasturing with the Mennonites. Some became rich with Mennonite help; some became poor. Then a severe epidemic among the sheep coupled with the drought of the mid-1850s led to the eviction of the Nogai, the first to leave the region, fleeing their generally poorer land. In 1859 the Nogai people numbered 35,000 in Molotschna. A year later there were only 100 Nogai remaining, the rest of the tribe part of a vast exodus to Turkey.[29] In this case, the Mennonites lost too. The Nogai land that the Mennonites hoped to use reverted to the state and was reassigned to Bulgarian colonists.

No wonder that as a child on the farm I remember wondering about "presences" I sensed on the land—not ghosts or spirits exactly, but rather the marks and imprints of those who anyone could see had lived on the land before we came. The small four-room house felt like it was inhabited only by my immediate family, but I often wondered who had stood on the worn wooden floors of the granary rooms in the tin shed where we played while my father worked. Small, completely wood-laid rooms, they smelled of grain and live presences. At night, I feared what I believed might be even older spirits, presences I sensed on the land.

Solo trips into the night frightened me as the moon shadowed the large Chinese elm trees around our house and the wind fluttered shadow figures on the ground before and behind me, figures who performed menacing dances round the tree trunks. How could the wind have hurled the tree swing at me as I ran past to the outhouse? Sometimes I tried to walk slowly and carefully into the impenetrable dark, allowing my eyes to adjust, hoping my ears would hear what I could not see and alert me in time to escape the dangers of the night.

The weird glare of the moon sometimes spooked the cows too. They shuffled and lowed uneasily in the lot just east of my dark path, stirring up clouds of dust I could smell; they rustled against one another with that head-butting breathiness cows offer to greet intruders. Just then the horse snorted and whinnied at the far end of the cow lot near the old windmill, creaking in the night breeze. I stopped dead still and held my breath, quieted my footsteps to listen, my heart pounding.

Suddenly, the horse's hooves clapped the sandy earth, pounding their way out of the familiar lot and fleeing north down the fenced lane. Why was the horse galloping into the night? Now I ran too—fast as I could go from the navy blue night into the yellow kitchen light where I breathlessly slammed the door behind me. My mother stood by the kitchen window, watching for my safe return. What did she fear in the dark?

I suspected that it was not animal. It was human. Sometimes today I remember my fear and recognize that we feared almost instinctively what white trespassers before us had feared: the eerie light of what became known as the Comanche moon. Our land just inside the southern edge of Seward County along with a part of the vast Texas and Oklahoma Panhandle lands had once been known and feared as Comancheria.

During the eighteenth and into the nineteenth century, the Comanches became legendary for their magnificent horsemanship and mobility on the plains and were able to keep incoming groups at bay. They burned newly formed settlements from Colorado to Mexico, from the western edges of New Mexico where they had first crossed into the Southwest, to Louisiana. A force to be reckoned with, they held off the intruders coming onto the plains from the east, decimated the early attempts at forming the Texas Republic, kept Spain and Mexico at bay, and ruled for a hundred years the entire High Plains desert area. They built an empire on the plains and left a legacy in the minds of those who would come after them.

The Spanish had early on taken note of the Comanches (or Numunu) coming into the southern plains from the west over the Sangre de Cristo Range in the early 1700s. Before long these Numunu were engaged in a full-scale battle to dislocate the Apache people. The new arrivals had been drawn to the plains by the bison, grass, and horses. There they also found guns. As new immigrants to the plains and excellent at adjusting to whatever they found, the Comanches were early opportunists. "By the 1710s, only a generation after obtaining their first horses, Comanches were lashing northern New Mexico with uncontainable mounted raids."[30] Infamous for their night raids "by the light of what was already widely referred to in Texas as a Comanche moon,"[31] the Comanche warriors knew the land so well that they were impossible to follow in the dark.

In the late 1950s my parents gave away the collection of arrowheads they had found while clearing our farm, a collection which could have told me something of the series of migrations of plains tribes who must have inhabited or ranged over the land we lived on even before the Comanches or the Cheyennes' hunting parties after them. I would have liked to know whether some in the collection were the tiny half to one inch points used by the Antelope People who lived at the Roy Smith site south of us. Or were they from the later Kiowa or Pawnee tribes who also would have crossed our land? The plains tribes who roved the land east of Liberal about 1300 AD would come to be known as the Pawnee. Much later, for a short time the Cheyennes were given reservation land which included the western fourth of Kansas, including the land still in our family.

When I realized that I had grown up on the northern edge of Comancheria, I sought out the few artifacts we still had, the carved and chipped flint arrowheads, knives, scrapers, and hairpins or door pins Mama had saved in her white Lucite box with Victorian blue leaf edging. A friend educated in arrowhead provenance took one glance at the collection lying on cotton in Mama's little white box and recognized most of them as coming from Palo Duro Canyon in the Texas Panhandle along the Canadian River, in the very heart of Comanche country. They are Alibates flint, like those found on the Roy Smith Ranch, red and white stone which resembles well-marbled meat and has been used for making tools at least since the Clovis Mammoth Hunter period 11,000 years ago.

My husband and I headed for Amarillo with the arrowheads, scrapers, and other knapped pieces my parents had unearthed to see the site from which natives in our area had brought this highly desired flint, the Alibates Flint Quarries National Monument. In "History and Lore of Alibates Flint," Wes Phillips notes that the flint is a secondary deposit in dolostone formed in shallow seas in the late Permian period 260 million years ago. I wanted to compare the marbled rainbow flint in Mama's box to that available in the Alibates Quarries near Amarillo—named, by the way, for cowhand Allie Bates, who worked for the rancher who once owned the entire area.

The only national monument in Texas, the 1,371 acres in Potter County known as Alibates Quarries is now managed by the National Park Service. Long prized for the ease with which it can be knapped into tools, Alibates flint in its various knapped or semi-knapped states has been carried throughout the Plains states and beyond for centuries. A strange, almost sacred feeling came over me as I walked on the chipped and knapped debris left so long ago or knelt to pick up a piece of flint left by someone who had used another rock as a hammer to chip away the piece I held.

Estimates suggest that there are a thousand pits for quarrying in an area ten miles square, and many more to be found just off the national park site. Native peoples came here and chipped out "blanks," partly worked smaller pieces they could more easily carry great distances for trade or to be worked later. They dug with tools made from bison tibias. Hammer stones, big sledge hammer rocks, were used to shear off large pieces of flint which would then have been broken with smaller hammer stones, pried with deer antlers from the outcrop, and shaped with sharp sticks or other materials.

A village near the quarries was inhabited by Plains Village Indians believed to be ancestors of the Pawnee or Wichita tribes. They once lived in rock slab houses akin to those we saw at the Roy Smith Ranch and farmed along the Canadian River—now dry—between the years 1150 and 1500. Perhaps even earlier peoples had first spotted these colorful rainbow pieces of flint in the river and learned that several feet under the caprock lay this excellent tool-making flint, harder than steel but more brittle and glass-like.

Later at the museum in Canyon, Texas, I saw the bison scapula's strong resemblance to the modern-day spade I used as a child to dig in the sand. In excavations at the 1874 Battle of Adobe Walls site in Hutchinson County, Texas, which we passed enroute to the quarry, gunflints of Alibates flint were found. Indeed, in a 1976 flintlock rifle demonstration, a participant using Alibates flint got 100 shots per flint as compared with 30 to 40 from English and French flints.[32] Amazing, the staying power of Alibates flint weaponry.

I am not expert enough to identify precisely the usages of the Alibates flint pieces in my mother's collection; however, I am especially intrigued with the one my friend John suspected could be a tent pin or a woman's hair pin. About four inches long and an inch wide in the middle at its widest, in an elongated diamond shape, this piece is absolutely smooth to my thumb's touch on its back side; it would serve perfectly to hold a knot of long hair in place. Though I can't be certain how it was used before it was dropped on our land, I want it to be a woman's hair pin. I want it to be the hair pin my mother found soon after her arrival on the land while she was digging the foundation for the dairy barn where she would spend her productive years milking Holstein cows with my father. Something in my mother must have recognized it for a tool or an adornment, something worth keeping.

That day my husband and I walked the site of the Alibates Quarries we remarked upon the stark and beautiful views, the harsh soil. The park ranger noted that it was as dry as he had ever known it to be, so dry that if it did not rain soon, he could not imagine what might happen to the local inhabitants. The drought. The so-diminished manmade Lake Meredith. The dry Canadian River—all unite us with the people of these plains, those who left here in the year 1350 and those of us who left the area east of Liberal.

I often remember the legacy of the Comanches as I drive the land around Liberal, still known as horse country. In the times before we arrived, if you wanted a horse, you needed to see a Comanche. No other tribe learned to breed, geld, and collect horses like the Comanches. A Comanche warrior might have one or two hundred horses, his chief 1,500 horses.[33] The amazing Comanche warriors were legendary for their ability to lie horizontally alongside their horse used as a shield, heels clamped over the horse's back, as the adept warrior fired arrows at the enemy. Of course, to keep and trade such huge numbers of horses, a tribe needed to control vast stretches of land. Eventually, with encroachment into their territory from all sides, the Comanches could no longer hold enough territory to sustain the horses they needed to maintain their hold on the land. Some version of that story on the land east of Liberal echoes throughout history: they could not hold the land.

Hauntings

I had written about the Comanche moon, my fear of the dark, and my sense of presences haunting the night before I encountered the concept of "haunting" as it is treated in Healing Haunted Histories. *I recognized intuitively the claims of those who talk about haunted landscapes. Chief Sealth, a Duwamish and Suquamish leader in Puget Sound during the mid-nineteenth century, is purported to have warned the white settler of their future: "At night, when the streets of your cities and villages shall be silent, and you think them deserted, they will throng with the returning hosts that once filled and still love this beautiful land. The white man will never be alone . . . " (quoted in* HHH, *37).*

As described by Enns and Myers,"[G]hosts inhabit both the geographies in which we settlers dwell . . . and the genealogies we carry in our bones" (39). Such hauntings are manifestations of abusive systems and violence that make themselves known and demand change. If we as settlers seek to deny their existence, it is at our own peril. Repression of such violence, denying our own unconscious, will eventually make us sick at heart. How then do we heal? We draw on the wisdom traditions of resistance and renewal, including our own Mennonite stories, hymns and treasured Scriptures. Enns and Myers argue that the work of faith communities is "to nurture the courage to peel the settler colonial onion seven layers down, fueled by the prophetic hope intoned by Malachi that a day is coming when the Creator will burn the works of injustice to their roots, 'until the sun of righteousness rises, with healing in its wings'" (41).

Believing that the settler hauntings they are describing are neither the stuff of horror movies nor the psychological disturbances of a person in need of therapy, Enns and Myers are drawn to the parable of the unclean spirits as presented in the biblical book of Luke to illuminate the kind of haunting and healing they discuss. In Luke's depiction, when an unclean spirit has been cast out of someone, it wanders, looking for a place to rest before it returns home. Finding its place swept and in order, it brings seven more spirits, more evil than itself. This passage is offered as a description of the work of decolonization—like peeling an onion, of a necessity digging into the inner layers of systemic injustice. Enns and Myers conclude that healing can only happen in the "shared struggle to turn our pathological personal and political history around, because it's killing all of us" (41).

I have always longed to know the stories of those who were on the land in Oklahoma before my people came. As a child listening to my parents pronounce the names of their childhood friends of Native ancestry in central Western Oklahoma, I remember desperately wishing to have a meaningful name like Heap of Birds. I repeated the words and imagined a pile of blackbirds, not dead, but flocking. How interesting! A heap of birds!

When Peace Chief Lawrence Hart spoke at Bethel College, I followed him back to the Cheyenne Cultural Center, desperate to know his people's story. I knew none of my own family history beyond that of my one set of grandparents who were alive, and I felt desperate to know of my ancestral pilgrimage. Isn't a zeal to know also a kind of haunting? My own experience is a deep internal, nagging sense of unresolved issues; sometimes I think of it as a trespasser's conscience. Surely, the root of healing is first understanding, that small nub that could grow to more individual wholeness.

Learning the stories serves as a corrective to my own sometimes poisonous mythologies. For years I felt proud to refer to the land on which I grew up as Oklahoma's "No Man's Land." I had created my own heroic mythology. This land no one else wanted was so poor only dispossessed ragtag Mennonites would take it. But they were such amazing farmers that even the Dust Bowl could not drive them off the land. Frugal and hardworking, they prospered on the land in the Oklahoma Panhandle. Smart not to irrigate their crops and suck the water from the Ogallalah, they knew to depend on God and Mother Earth for their needs.

There are other ways to understand "No Man's Land" which I began exploring after I read Eula Biss' description of her experience of No Man's Land in Chicago. The designation goes back to the Old English in the 1300s, "nanesmaneslande" *used during the bubonic plague to reference the pestilent land between parishes where mass burials occurred and where no living person wanted to go. The term carries as well the notion of wilderness, or forbidden ground used during warfare for the unoccupied area between warring armies. Land unoccupied, uninhabited. The central sin of the settler across this country was to assume that the land they found was vacant, for the taking if it was not being farmed. Of course, the land we live on is haunted.*

The Comanche origin myth goes something like this: the Great Spirit of the Sun went to the four corners of Earth to collect swirls of dust to create the People. Formed from the earth, they had the powerful strength of mighty storms. Unfortunately, a shape-shifting evil power was also created and began to torment the People. The Great Spirit took pity and cast the evil power into a bottomless pit. But the Evil One took refuge in the fangs and stingers of all poisonous creatures; so it continues to harm the People.

Part III

Spring Equinox

Easter. Resurrection. Ancient symbols of rabbits and eggs. We boiled and dyed our eggs pale green and yellow and lavender and laid them on the springy artificial green grass we reused each Easter in our baskets. Mama sewed new Easter dresses in pastel colors for us to wear Easter Sunday, and we put away our dark shoes in favor of lighter colors.

Long before Easter we had begun to walk the green wheat fields with our father while he reminded us that the wheat needed to be tall enough for a rabbit to hide in by Easter. It wasn't really about hiding rabbits or Easter eggs, but rather my father's gauge to measure how his wheat crop had weathered the winter.

Winter is long for a dairy farmer who struggles with freezing conditions in the dairy barn, cows' frozen teats, the numerous inconveniences and annoyances that cold weather delivers, including the occasional snow blizzard which can be fatal for cattle. During the infamous blizzard of 1957 my parents could not get to the dairy barn through the snow drifts between the house and barn for a couple of days to milk our cows. The early thawing and new life of spring were both a relief and reason to rejoice. On Easter morning we rose early for sunrise services to announce the resurrection (and see the light). Later, the church sanctuary would be filled with a profusion of lilies and fragrance and sunshine and visiting kith and kin.

The Judeo-Christian narrative accompanied the natural rhythms of the earth: Christ's death and resurrection was emblematic of the seed that must die to come back to life.

Many years later, baking Easter Paschka breads with my mother to take to the neighbors, I read to her an old Plaut Dietsch lullaby I had found in an old cookbook, an Easter rhyme; she immediately remembered her mother reciting it:

> *Scheckel, scheckel, schei-ja*
> *Oostren et wie Ei-ja*
> *Pinksten et wie witte brot*
> *Stoaw wie nicht, den woa wie groot.*
> English:
> Swing, rock, rock, shush.
> At Easter time we eat eggs,
> At Pentecost we eat white bread,
> If we don't die, then we'll grow up.

"This is a lullaby?" I asked my mother. She had never used this rhyme with her own children. "If we don't die, then we'll grow up?" I was horrified. Mama smiled and noted that it doesn't quite translate. Such ritual sayings marking the seasons—whether they were conscious or unconscious—reinforced the understanding that our lives depend upon the earth. We could die; to live will take hard work to grow our crops, as well as luck and God's providence.

Designed to keep us humble about the tenuousness of our survival, the celebrations of the changing seasons also taught us to be joyful for what we had been able to grow. No matter how pitiful the crops we had produced, there was optimism for the future: this year's wheat would be better than last year's. The cycle begins again. You get another chance. We adorned ourselves in anticipation—perhaps not so unlike the ancient Puebloans who painted their bodies on feast days to make their sacrifices to the earth.

Except for an occasional Easter holiday, Daddy's brothers and sisters rarely visited us *en masse* at our farm. The women got the beds in our four-room house and the men lay packed like a layer of sardines in a can, elbow to elbow on quilts on the floor. Perhaps our relatives had waited until the weather had begun to mellow before they made the four-hour drive to our farm. Easter Sunday afternoon after we children had found the Easter eggs our parents hid for us, it was the children's turn to hide eggs for our parents.

This second hunt was our favorite part of Easter weekend when our parents turned into silly children. My father's siblings, thrilled to be together, became once again their childhood selves during the Easter egg hunt; feeling the end of a hard winter, they cavorted and chased each other around our yard and into the wheat fields like spring-crazed baby calves high tailing it, elbowing one another to get an egg they spied, teasing and laughing and falling through the cellar door. They looked reborn! Easter was resurrection in many ways.

INTERLUDE

MIGRATIONS

my ancestors surround me
like walls of a canyon
quiet
stone hard
their ideas drift over me
like breezes at sunset. . . .
—from Harvey Ellis, "ancestors" (used by permission)

My family's arrival on the land east of Liberal was the culmination of centuries of migrations by our Mennonite farmer forbears. There was a certain inevitability about my parents leaving their city jobs as newlyweds to seek the autonomy of a farming lifestyle. As a child I believed that everyone who mattered farmed. In my church everyone farmed and lived in the country.

I had no ability then to think about the impact of farming on human life or the way agriculture had changed everything in the human migration story. I didn't know that thousands of years ago agriculture as a way of life began settling people around the world, first in the Middle East and Central Mexico, then in China and the South American highlands, later in North America. They quit roaming, hunting and gathering according to the hunting season; they organized their lives around planting, tending and storing crops, and caring for the animals they began to domesticate.

Human populations naturally increased after settlement as daily life became less hazardous and more sustainable.

Families became larger because they no longer needed to carry babies on their backs for long distances. Both my maternal and paternal farming grandparents' had seven children in their families. Agricultural villages planted more crops, domesticated more and more animals, and because they no longer followed the roving food supply soon depleted the nearest supply of game or berries.

Settled now, they made pottery they would never have thought to carry in their earlier need to travel light. Where before they had used baskets woven of whatever reed or grass they found alongside temporary shelters, these settling ancestors cooked and stored larger quantities of food in granaries. Farming naturally leads to laying up stores of food for more and more people whose very existence depends upon hard work, so a work ethic becomes the paramount value in a farming culture. Settlement leads to villages organized according to work roles, specialization and better technology, accumulation of wealth, and religion.

The mythologist Joseph Campbell has highlighted the influence of early agriculture nine thousand years ago on religious practice in the Near East and Old Europe, especially pointing to the resurrection motif he found in the belief in the Goddess and the child who dies and is resurrected, springing from the Goddess and going back and resting with her. This mythology carried through Mesopotamia and Egypt and eventually found its way into Christianity: resurrection of the Christ figure symbolic of the buried seed which springs from earth to provide new life.

Centuries later, Mennonites would carry on these deep primitive and historical roots. Such influences were expressed in both their farming traditions and in their faith as they migrated across Europe into Russia and to the U.S.

From the Netherlands

My Mennonite forbears in the 1500s, a persecuted lot of farmers, often had to learn to live on the worst soils because their Anabaptist beliefs put them at odds with the State. On the run, fleeing those who might imprison them or sometimes demand their very lives, the early Anabaptists took whatever quarter they could get. Driven from place to place by their nonconforming beliefs, fleeing the dictums of the state they could no longer tolerate, they had to learn to live on the lands of those who would have them.

When the Anabaptists fled the Netherlands to places like Moravia and Prussia, where my farming ancestors lived—to land wherever some duke might tolerate them, they used their knowledge to drain swamps or otherwise improve the land. Eventually, they gained a reputation for being good farmers, willing to do the work others might not in creative or experimental ways. They often left one area as a church group or a group of families to move where they could live quietly. Often they were not allowed to live among established communities and were banished to live outside the city walls.

This was true of my Mennonite ancestors near Danzig. Because their young pacifist men refused to fight to help protect the city, they were forced to live outside the city wall where they could not proselytize, but could thrive on their quiet farm operations and bring their trade to the city.

I trace my family's migrations back to the time of our denomination's namesake, the Dutch Catholic priest, Menno Simons, born January 1496 at about the same time that Columbus landed in what would become the Americas. The headstrong priest Menno read and interpreted Scripture for himself, whereupon he began to doubt Catholic doctrine. For example, he questioned transubstantiation—whether the sacraments he was dispensing literally became the blood and flesh of Christ.

In 1536 Menno joined the revolutionary Anabaptists who had renounced infant baptism, went underground, and eventually became the leader of a movement which preached believers' baptism, putting him at odds with the state practice of infant baptism. As a hunted man, he fled with his followers from the Netherlands to the Lower Rhine region, Prussia, where, included among the Anabaptist groups and churches he helped to shepherd was the Mennonite church at Danzig.

To Moravia

During the early 1600s when Coronado was marching up from the Gulf of Mexico and across the land where I would eventually spend my childhood, my tenth great-grandparents were in Moravia. Both my maternal and paternal lineages cite Moravian ancestors born during the late 1500s and early 1600s. When Catholic authorities began intense persecution of the Dutch Anabaptists, they spread eastward in large numbers to the Tyrol and Moravia, a region of the Czech Republic located between Bohemia on the west and the Carpathians on the east; the chief town is Brno.

They found refuge on the lands of the tolerant princes of Moravia in an area which became a safe haven for persecuted Anabaptists during the last half of the sixteenth century; they left for Prussia in the 1630s.

To Prussia

My ancestor Tobias Hans Schellenberg, born in Moravia in 1603, migrated to Prussia from Moravia in 1634 during a time of persecution against the Anabaptists, known as Moravian Brethren. Church records show that Tobias Schellenberg came to Schoensee, Culm, Prussia on the banks of the Vistula River where a group of Mennonites had fled persecution and set up farming communities years earlier in 1553.

City magistrates at Culm apparently were good landlords to the Mennonites, and they prospered. The migrants to Prussia also had the advantage of the security net their faith community offered them. They stayed in contact with Mennonite elders in the Netherlands who visited them in their new Prussian homes, offering support. Some of these Mennonite villages had begun as early as 1540, when land was first sold to the Mennonites coming from the Netherlands at almost exactly the same time that Coronado was marching across Kansas.

In 2016 we visited Poland with a Mennonite heritage tour to see the West Prussian homeland of my farming ancestors for 250 years—and received our list of ancestors. Never having seen a genealogy of my kin, it was an amazing surprise: 67 pages listing nearly 300 names going back as far as my tenth great-grandparents. I was stunned. I scrolled through page after page in total disbelief. Beyond my parents' grandparents I had never heard of any of them. Remarkably, however, the surnames assembled from the Mennonite database "Grandma" by the leader of our tour to schedule visits to our ancestral village sites were almost uniformly recognizable to me: Richerts, Regiers, Schmidts, Goertzens, Schmidts, Bullers, Pankratzes.

Clearly, these were my people back into the 1600s—excluding my father's Lutheran ancestors. However, my father's Lutheran grandfather, whose family left Hamburg in 1874 when he was a boy of nine to come to the U.S. on the Teutonia with a shipload of Mennonites, joined the Mennonite church when they arrived in Kansas, and this great-grandfather married a Mennonite woman. Hers were the oldest ancestral records in my list. I would learn as we visited the Mennonite villages in Poland that Lutheran villages were frequently located nearby; there were lots of Mennonite/Lutheran marriages.

I found it powerful and moving to look through my *Ahnantafel*, literally, "table of names" (though Hitler and the Third Reich desecrated the word *Ahnantafel* with their emphasis on Aryan pedigree). I held in my hands my very own "great cloud of witnesses." I will also admit to some chagrin when I realized how provincial these Mennonite communities must have been for me to recognize, centuries later, all of the surnames. Of course, Mennonites chose to live in community, and often intermarried with other Mennonites; thus, the family names remain even today a part of the Mennonite diaspora scattered in South America, Mexico, Canada, the U.S.A, and those resettled in Germany.

There is little story line with most of the genealogical entries, a bit like the begats in the Bible: birth, marriage, children, migration, death, whatever might be known. Sometimes there occurs an interesting tidbit: a certain Ratzlaff, no first name given, is the first Ratzlaff to join the Mennonites. He was a Swedish soldier who, upon hearing a sermon, stuck his sword into a hedge post, breaking it in half (or so the story goes), and subsequently joined the Mennonites through baptism. We know who he married—the daughter of Aeltester Voth of Kulm. There are claims he went to Holland to be baptized. However, "there is no evidence."

Reading this account, I realized how names influence my sense of peoplehood. My father used to call me "Minna Ratzlaff," teasing me with a heavy German accent, referencing an apparently somewhat flamboyant tall woman in his extended family who had red hair and wore hats, an older relative of his. We children thought "rat's laugh" a totally ridiculous name.

If the surnames are familiar, the women's names are bizarre to my ear—Alcke, Ancke, Efcke, Elscke, Lencke, Liscke, Trincke, Trudcke, Sarcke, or, the strangest to my ear, Buschke. Yet, even these names hold a certain wry familiarity. "Buschke" sounds distinctly like some ironic name my circle of girl cousins or aunts might have pinned on one another in play when we denigrated one another with German-sounding epithets in play and in mockery of our German past, mixing German words into English: "Oh, Buschke, tie your hair in a schups (bun), put on your doak (head scarf) and come sit here and peel potatoes!" So, even the unfamiliar here rings somehow familiar—like it might have come from some deep unconscious reservoir.

The Polish countryside is lush in June, as the bus lurches along filled with a group of Mennonite-connected tourists. "If there is a clear spot in the sky the size of a Dutchman's pants, it will be a clear day," says Dorothy Bergman, a delightful Canadian woman and fellow traveler to Polish heritage sites, quoting her grandmother. Farmers the world over care about the weather; the oral history we inherit helps to record our ancestry back to the Netherlands. We name the crops we see out the bus windows: rapeseed (canola), rye, barley, something that looks vaguely like wheat. I feel at home as I do driving through farmland anywhere even as we drive alongside the dikes built by Mennonites. Village names are in Polish, but the Polish government welcomes us with historical signage marking our ancestral villages and cemeteries, even if one needs a good guide to find them.

The Polish landscape and its creatures pull me in. One of the first Polish words we learn is *bocian* for stork, as Eva our

Polish guide points excitedly at every example of the massive stick nests atop electrical highline poles or chimneys, sometimes including the white stork visibly perched on the edge of the nest. Polish folk legends have a stork story that reminds me of the biblical Noah:

> Once upon a time, frogs, snakes, lizards, and others of their ilk were multiplying excessively on earth. Because they were making a nuisance of themselves, God gathered them up in to a sack and, calling man, told him to empty the sack into the sea. Man—feeble and weak creature that he is—was unable to contain his curiosity and untied the sack along the way to see what it was he was carrying.
>
> All of the creatures slipped out and scattered, hiding themselves lest they be gatheered up again. As punishment God turned man into a stork, a *bocian*, and condemned him to hunt the creatures for the rest of his life.[34]

The legend resembles Noah's story but also the plight of Adam and Eve, condemned with the rest of earth's farmers to hoe weeds the rest of their lives! At least Adam and Eve were allowed to remain human sinners. Clearly, for Poles the human and the stork, whose life span of seventy years is equivalent to that of a human's, are kindred spirits; when the stork returns in the spring from its migration to South Egypt, it is to be welcomed to nest on the chimney to bring good luck.

Eva, who has served for many years as the Polish guide for these Mennonite heritage tours, knows a great deal about the sites we visited and has many of her own tour routines by the time we are part of her entourage. Like passing candy. She seems almost obsessed with candy, begging us to try another Polish type of candy every day as she makes her way around the bus. I always oblige her.

My father also was obsessed with candy. Is this too a Polish tradition we Mennonites carry on? Why do I so love the familiarity of candy? Is it my selfish need to understand my own sugar addiction? Genealogy, that magnet of the "familiar," is a strange and perverse and lovely and even insidious attraction.

The Mennonite name *Wedel* is still the brand name of a great Polish chocolate maker. We buy up candy bars. Purportedly a Mennonite-founded distillery is still making Goldwasser, a vodka with real gold flecks in it that we had trouble getting through customs, not because the customs officials were reluctant, but because they were curious: "Are these real gold flecks floating in this bottle? Is this stuff expensive?"

Mennonites farmed the Vistula area for 400 years, 1534-1945. Their claim to fame was land drainage, a multi-generational project. First, they drained enough to produce meadows and pasture with excellent grass which didn't suffer when flooded. They became good at dairying and bred a milk cow known as the "Milk Boat," a Werder cow. The Holstein would not come into this area until 1852.

The Mennonites also developed a flat funnel-like earthenware cream separator and churn and learned the art of making cheese. The women's claim to fame was their cheese making and gardening; Tilsit cheese, still popular today, came from the Mennonites. The Mennonite farmsteads were also known for fabulous gardens, no doubt planted, maintained, and preserved by the women, if my grandmother was any indicator.

We are told that windmills once dotted the landscape in certain places in the Vistula Delta. In the Weichsel River Delta near Gdansk there were 200 windmills to help drain excess water where the marshlands were flood prone. There may also have been as many as 100 grain mills. In recent years the last beautifully intact Mennonite windmill burned. Our tour leader wept when he heard the news.

My sister and I had at least one almost mystical ancestral experience on our Polish journey. Our records told us we had people at what was once *Przechowka* (sounds like chuh-huff-ka). A Polish factory was now using the old village site for a dump, but what remained distinct and symbolic for our group was the prototypical, straight-as-an-arrow road running down the center of the old Prussian Mennonite village—surrounded on either side with the dense foliage and growth of neglect. We clambered out of the bus to walk on and photograph the old road.

All of us began to pick up and pocket fragments of rocks, not because they were old, but because they were from a site that was once Przechowka. We photographed the wild flowers alongside the road; we noted the green growth over what our heritage guide Alan told us was once the location of the village cemetery. In fact, Alan remembered an earlier group he brought to this village site years ago when they could still walk among the cemetery stones, now completely covered in tangled forest growth. He smiled to remember that eager earlier group who were greeted with stinging nettles when they rushed into the growth.

We stayed on the road, grateful for the warning. The earlier group who walked among the cemetery stones had found the marker of a Tobias Sparling born in the 1600s, and, seeing that the cemetery was going to ruin, decided that they should retrieve and carry out this stone. They heaved it onto the bus, now trying to decide what to do with it. "I'll show it to you," Alan told us when we are back on the bus. "We ended up placing it in the much better kept cemetery at Heubuden which we will be visiting."

Back on the bus, my sister Vicki hails me. She had listened to the story of the moved tomb marker. "Don't we have a Sparling?" she asks. We do. Tobias Sparling, born "about 1675," part of our maternal Grandpa Kliewer's clan. Alan remembers the dates for the Tobias Sparling whose stone they removed as 1690-1757. If our Tobias was born in 1690, he likely wasn't married in 1696 as our records show; yet, would there have been two Tobias Sparlings so close in age at the same place? A father and son? Could the dates be off? "Take him," Alan says. "Claim him as your Tobias Sparling. I'll show you the stone." We do claim him, and anxiously await our chance to see his stone at Heubuden. Genealogy is imprecise, we know, as it pieces together the fragments of others' memories and chronicles. And we all wish so desperately to claim an ancestor.

To Russia

My great-grandparents (not including the Hinzes) were all born in South Russia in various villages of the Molotschna colony. So, too, my great-great-grandparents, born in the earlier 1800s from around 1820 to the 1840s. Most of my third great-grandparents were still born in Prussia before the migrations to South Russia. Even my father's people, born near Danzig according to the family Bible, my Lutheran great-great paternal grandparents, migrated to Russia where they joined the flood of Mennonite farmers who came from Russia to the U.S. in 1874, now centuries after Menno Simons' death in 1561.

My father's great-grandfather's family was among the very few Lutherans in the group of mostly Molotschna colony Mennonites who came from Russia on the Teutonia. My mother's people were among those from the majority Alexanderwohl colony in Russia. When Catherine the Great decided to develop agriculture on the steppe in South Russia, the Mennonites around Danzig, especially, under Prussian control after Poland was partitioned, were eager for more autonomy and were lured by Russian promises of free land, expenses, freedom from taxation, and exemption from military service.

By 1869 there were 40,000 German Russian Mennonites, half of whom lived in Alexanderwohl, one of two large South Russian settlements ninety miles from Berdiansk. There they flourished as their hosts had hoped, planting orchards, raising cattle and sheep, and farming. Their wheat production was a significant Russian export. However, things had changed in Russia by 1874 when my ancestors left their prosperous farming villages. Their sons were by then subject to Russian military conscription. The Russian government hoped to equalize land holdings, treating the wealthy Mennonite landowners just like the Russian peasantry. Their separate schools were being threatened. Thousands of Russian German Mennonites elected to move to the U.S.

and on to Kansas and Oklahoma

By ship and then by railroad, my ancestors first arrived in the U.S. and took the railroad to Kansas in 1874. Later, some would make their way to central western Oklahoma where my parents grew up. Historians have noted the irony of the peace-loving Mennonite farmers coming to settle in Kansas—known in the 1870s for its lawlessness and grasshopper plagues. The amazing advance of the railroad across the state was what brought them.

Promoters were promised eight and one-half million acres of Kansas prairie by the U.S. Congress if they would agree to build a railroad. By 1872 the Kansas Pacific and the Atchison, Topeka, and Santa Fe had accomplished the task and earned the right to claim seven million acres in alternate sections twenty miles on both sides of their right of way.[35] The railroad civilized the rowdiest cow towns, including Newton, in the center of the state, the town which would become the Midwest center for Mennonites.

Both my maternal and paternal great-great-grandparents sailed over the Atlantic from Hamburg on the Teutonia in 1874. My father's Lutheran family joined the Mennonites in Kansas at Hoffnungsau ("field of hope") Mennonite Church near Buhler, one of the earliest Kansas Mennonite churches. However, both my maternal and paternal forbears eventually went south into Oklahoma, where they farmed and ranched on the red soil near a town named Korn, later anglicized to Corn.

My parents were too busy as part of the post-WWII "Greatest Generation" building a life to look back at their ancestry. Or, perhaps they simply took their histories for granted. They never spoke of ancestors. Their first schooling had been in German. English, and German or *Plaut Dietsch* were interchangeable in their homes, and they spoke the old language to their grandparents. As an adult I had to seek out for myself the details of the long journey which brought my parents to southwest Kansas.

Migration journals written by both my father's maternal great-grandfather Ediger and my mother's paternal great-grandmother Froese leave scant details. Grandfather Ediger left Schardau in Russia and traveled by railroad, carefully recording the hymns they sang together as they joined with other Mennonites along the route, hymns like "In Jesus' Name We Journey On" or "God Is Our Leader."

He also records what he notices as a farmer—for example, at Kustino, journeying out of Russia: "Here were many fields of carrots" or eventually, leaving Berlin, "The cattle were all black and white. . . They did not have such cattle in Russia." How could he know that his great-grandson, my father, and his wife would make their living for several decades milking this very breed of cows in the "new land"? When the ship was nearing the American shore, Grandpa Heinrich Ediger records that they sang, after seasickness and storm, "Dear Jesus, We Are Here."

This grandfather Ediger writes that though his wife and children enjoyed the 1500 mile train ride over the land, he did not. "Closer to Michigan the land was better and each farmer had his land fenced. . . . Very good fruit trees, lots of apples. . . . At 10:30 we went over the Missouri River. We bought a bread and watermelon in Kansas." And so his farm life began: "October 6. We got on our land. We built us a house. It was 20 foot long and 16 foot wide. We plowed 5 acres up. We planted 4 acres in rye. We did much plowing of land."[36]

Between six and seven thousand Russian Germans arrived in Kansas in 1874. Grandpa Heinrich Ediger would remain a Kansan, but his son Jacob Henry Ediger and his wife would relocate to Oklahoma in the early 1900s, and try to homestead in Littlefield, Texas, before settling on a farm near Corn.

We have been unable to find the central Kansas grave of Christoph Hinz who brought his Lutheran family to this country, though the census listing his first farmstead is intriguing. The 1885 Census in Harvey County in Kansas where he first farmed was difficult to find given that the census taker's record had somehow transcribed "H-I-N-Z" in cursive as "K-I-N-G," but we eventually found all the family members' given names and confirmed the agricultural schedule for Christoph Hinz and Johanna Freitag Hinz as follows: 160 acres of land and 840 rods of hedge fence. They had planted 120 acres of winter wheat, 15 acres of corn, and one acre of Irish potatoes. They had 300 bushels of corn on hand. There were ten acres of prairie under fence. They claimed 8 tons of tame hay, 12 tons of prairie hay cut in 1884. The family had sold $10 in chickens and eggs, had made 100 pounds of butter, and owned nine horses, 3 milk cows and 2 other cattle. They also had 11 pigs, had lost 2 horses, 1 cow and 1 pig, and had slaughtered $60 worth of animals. On the farmstead were 200 peach trees, 2 apple trees, 114 grape vines, and 1 dog.

Their agricultural holdings sound quite prosperous to me, but they would lose this land, I suspect in a drought, along what is still known today to local Mennonites as "Dutch Avenue" between Hesston and Buhler, Kansas. When I finally located the land they had homesteaded to explore it for myself, I found it completely converted to farm land—no sign of a homestead, farm house or outbuildings. Though I have found and visited my great great-grandmother Johanna Freitag Hinz's grave in the cemetery on the edge of Moundridge in central Kansas, I have been unable to find the grave of this first grandfather in this country, though we know that when he died he had a harness shop in Burrton. Would they have buried him under an apple tree, or has his marker simply disappeared?

A woman's version of migrating to this country comes from my mother's great-grandmother, Elizabeth Froese, from the Molotschna colony and the village of Landskron. The journal notes are brief, understandably, given that Elizabeth was a young mother of a nearly one-year-old son when she and her young husband came on the *Teutonia* in 1874. She records that it took eighteen and one-half days to cross the ocean and ten interminable weeks to get to McPherson County in Kansas by train, with lots of layovers. Her concerns are geared more toward housing than the lay of the land. Where would they live? She tells of staying in immigrant housing the railroad provided and writes almost casually—at least in translation—that the immigration records changed her son's birth date: in Russia he was born on October 4 but "in America it was October 16."

This grandmother Froese continues to record births, illnesses, and deaths. She is desperately ill with "fever" after her daughter's birth six years after their arrival and remained desperately ill through both her daughter's and husband's deaths with the same fever. Her journal reads about the little girl: "She is buried in Kansas somewhere." With her second husband (it seems so matter of fact), she moved with other Kansas Mennonites by covered wagon to Oklahoma land opened for homesteading in 1894. They lived in their covered wagon until they could build a dugout: "They plowed up the prairie and with the bricks they built the walls of the dugout. It was partly underground."

No wonder that I still recall my first thought when I heard Cheyenne Peace Chief and Mennonite minister Lawrence

Hart's 1995 commencement address at Bethel College. In that speech he said that he hoped, if it had not been plowed, that he could find the 1868 site of the Southern Cheyenne annual Renewal of the Earth ceremony during which they would have raised a cottonwood tree to the heavens as an *axis mundi* somewhere nearby. My heart sank. I knew immediately that he would never find it, that my people had, indeed, plowed the land. The refrain in both my grandparents' ancestral journals was always: "We plowed up the prairie. . . ."

East of Liberal

Growing up in the '50s, I always thought of myself as an immigrant or outsider, not really an American like the other kids in my school. We called them "English" kids as opposed to us "Germans." I was not aware that a true immigrant classification would necessitate that I be in the first or second generation. Perhaps because I was Mennonite I felt my strangeness and non-status quo belief system even if it was not obvious in my lifestyle. I tried not to look or dress like I came from a different tradition, but often I felt like an outsider at school where there weren't other Mennonites from my church.

Probably this is because a central tenet of the faith even liberal Mennonites teach their children is that they should live in the world, but their behavior and beliefs should not be "of the world." To a kid who doesn't understand Mennonite theological concepts like pacifism, it simply felt like we spent a lot more time in church than our school friends did, and we took our lifestyle cues from the church community. That meant we didn't dance, drink, have television, or socialize that much with neighbors other than church friends or family.

My parents would hardly have thought about, remembered, maybe even known of their grandparents' fairly recent Kansas history when they crossed the Oklahoma border into Kansas to settle on their new property in 1950. They were returning to settle in the state where their great-grandparents had arrived on the train in 1874 and set out to farm.

Mama always said that she never intended to live on a farm like she had as a child growing up near Corn during the Depression years. I expect there were lots of Depression-era kids who promised themselves a different life. In their families of seven children, Mama and Daddy were both near the middle of their broods. Mama was the third girl born into her family and Daddy the third from the youngest of the seven in his family. They fell in love in high school at Corn. Mama was a smart and pretty twirler, Daddy an athlete, at two inches over six feet, and class president. After Daddy went into the army to support his widowed mother, Mama, as Daddy's fiancée, worked in Clinton waiting tables, then in a small grocery, and eventually in a bank, waiting for Daddy to return.

A citizen of Tom Brokaw's "greatest generation" of men who served in World War II, my father declared as a Mennonite noncombatant and was assigned to be an interpreter who would translate German to English. He served in the last phases of the war as he helped to interrogate captured German soldiers. Mama still has the picture frame a German prisoner carved, painted, and decorated with red roses for Daddy, and it still holds Mama's engagement picture, flashing her happy smile on a face framed by beautiful waves in her hair.

It took me years to figure out why my daddy had been in the army when most Mennonites chose alternative service. His cousin told me long after my father's death that when his father Ferd died suddenly when Daddy was seventeen, his mother Sara was left with two younger children and no means of support. Daddy's uncles advised him to go into the military as a noncombatant after he finished high school; the army would send his mother money to support his younger siblings, aged eleven and fourteen.

Sometimes apparently belief gives way to necessity, though I learned later that there were many Oklahoma Mennonites whose patriotism called them to put country over religious principle for the sake of standing up to Hitler. It has always interested me that my daddy's grandfather, the old German patriarch of the family, who had come to this country at age nine, was writing letters to his German relatives or friends and ordering German medicines from Germany during those WWII years.

Perhaps this is why these kinds of decisions never got discussed in my hearing. How do you sort out the theological and the patriotic versus the economic need and offer a sensible explanation? And then, often in my daddy's family when I asked these kinds of questions, they merely shrugged, as if to say, how do you second guess all of this? We did what we had to do.

The Army sent my father to Heidelberg University to brush up on his German dialect and study German culture; thereafter he rode with the "army brass" as an interpreter, coming behind the Battle of the Bulge and helping in the liberation of death camps. These are not things my father spoke about, except that once he mentioned "train carloads" at the death camps as his eyes wandered far away into some space in time as if to avoid the unspeakable. He always promised that once he sold his dairy cows, he would take Mama to visit Heidelberg where he had studied. Hearing that, I often fingered tenderly the post card he had sent her kept among his military memorabilia: *Ich hab mein hertz in Heidelberg verloren.*

Mama would have to visit Heidelberg without him, however, in the company of her widowed friends and retired age mates, many years after Daddy was gone. My husband and I too sought to find some lingering presence of my father's stay in Heidelberg, years after he had died. We found plenty of name plates with the name of Hinz, and we found the soft and light sugar cookie called an *Americanisher* which the baker told us was named for the American soldiers who loved them. We also learned that Heidelberg was one German city that had remained intact; Americans loved Heidelberg and preserved the city from bomb attacks, we were told.

Nearing the end of his term of service in the Army, Daddy was called home when his mother, not yet age fifty, lay dying, her family believed, "of a broken heart." The family story goes that Daddy could not disclose his location in letters during those secretive WWII days. His family tried to send telegrams notifying him that his mother was dying; however, he was typically moved to a new location before they could reach him. Back in Oklahoma, the doctors shook their heads at my grandmother Sara's terrible breathing and told the children each day that she would die that night. But my father's mother declared that she would not die until her son returned to say goodbye. And that is how it happened. My father finally did receive the news of his mother's impending death, and the Army "rushed" him home on a Swedish trawler. As he took his dying mother into his arms, she asked him to take care of the younger children and died shortly thereafter.

Stories like that one always encouraged me to believe my father a bit of a hero. He was the tallest brother, the one in the family who had traveled, the one who had come to love culture in Germany, who had listened to "Ave Maria" in the great cathedrals; I thought he was the brother especially devoted to education and the exploration of the Christian lifestyle. His bearing and demeanor were military and dignified; posture was important to Daddy. When as a young girl I grew too fast and towered over everyone in my class, I began to slump, round my shoulders, and try to either fit in or disappear. Daddy threatened to build a wooden crosspiece to place on my back and straighten my posture after he saw me slink down the aisle to the front of the church. I threw back my shoulders and raised my head.

I admired my father's posture and loved to watch him unfold his tall frame as he removed himself from the driver's seat in our long sleek green 1960s Mercury after we had driven into Liberal from Sunday morning church services in our Oklahoma Panhandle church. We didn't need groceries; Mama would have prepared a sumptuous Sunday dinner of roast beef, carrots, potatoes, and onions which would be baking in the oven at home, to be served with the *zwiebach* (homemade double-decker dinner rolls) and pies she had baked on Saturday. Rather, Daddy was getting out of the car so that he could get the Sunday papers to read that afternoon—*The Daily Oklahoman* out of Oklahoma City and usually also the *Hutchinson News* which covered western Kansas.

Daddy marched into the Jack and Jill on the southeast end of Liberal in his black suit, white shirt, tie and dress shoes, to return smiling, his boisterous, gregarious self, still engaged in friendly banter with the clerk or other customers; my daddy loved people and they loved him. He opened the car door with all the swagger of a returning prince and tossed his treasured papers onto the front seat where Mama took them to scan the headlines, and then he threw in the candy, a sack of maple nut goodies or peanut clusters, something for his sweet tooth and afternoon snacking while reading the paper on this day of rest.

Sunday rituals were key to breaking the life-sucking sameness of operating a dairy farm. We went to church to sing about God's good earth and our role as stewards; we dressed up to remember that we were more than lowly ploughmen; we sang in four-part harmony with our fellow worshippers, needing the aesthetic, the spiritual sustenance, and the community; we played the piano and organ; we studied the Bible and stretched our intellectual capacities with hard theological questions; we believed we were honing our skills to become better humans and more Christlike; we socialized with the likeminded in our congregation and heard how their crops were doing; we commiserated with those whose lives were scarred by rain or wind or fire; we ate well to tell ourselves that we were prospering! Unfortunately, we still had to milk the cows twice a day, a rhythm that never stopped. Mama and Daddy raced out to make short order of the task Sunday morning and evening, in order that they might focus—insofar as possible—on the altered rhythm of the Sabbath.

Vacant Land

As a farming people we have inherited a conqueror's mindset and a conqueror's privilege established centuries before we used it—even though we spend our lives talking about humility. Somewhere within the set of understandings we call today the Doctrine of Discovery is the notion of terra nullius, *Latin for "uninhabited earth," translated by the sovereign powers who established it to mean that unless another country got there first, the colonizer could freely take the land for his own sovereign, land considered vacant unless inhabited by white "discoverers." Native peoples' long sojourn on the land was ignored.*

Lay that concept alongside the notion of "no man's land." Whether by disease, legal concept, treaty, invisibility, the myth that found Natives to be sub-human and pagan, outright land theft, forced removal—what resulted on the land we now inhabit was genocide for Native peoples. Ironically, the same year that my grandparents came to Kansas to claim land to farm in 1874, the last great battle against the tribes occurred at Adobe Walls in Texas, removing tribal people from the land our people then took up to farm. Thus, my ancestors could allow themselves to believe the land was vacant.

I have thought often about why my ancestors left their prosperous homes in Russia, but not enough about the land they came to and settled on in Kansas and Oklahoma. The stock answer I received about how they became "landed" in this country was that the railroad provided land, that other Mennonites had preceded them and "made the way." We have often been told how these ancestors stayed in immigrant houses until they could go to the land where they lived hard lives in dugouts until they could break the soil and begin to farm.

When I reread my ancestors' journals, I realized that there were notable details I had missed or ignored in earlier readings. For example, when they arrived in Topeka after their railroad journey across the country in 1874, they were shown land in the nearby area, making excursions to survey land in nearby Kansas settlements, including Council Grove, Florence, Marion, Bruderthal, Ellinwood, Great Bend, Hutchinson, Halstead, and eventually, Newton.

Some years earlier, in 1848 the Kansa people had been relocated from their longtime sojourn along the Kansas River in northeast Kansas and what is today Topeka, to what they were promised would be a permanent home, a 256,000 acre reserve including the upper Neosho River Valley and Council Grove, the white settlement and last stopping point on the Santa Fe Trail between Independence and Santa Fe.

No one today disputes that this relocation of the Kansa people to this site was a terrible mistake. The forced removal from their longtime homeland along the Kansas River led to what are now described as their darkest years in this state. At Council Grove the Methodists built a school, but the Kansa children rarely attended. They were not provided an agent as they had been promised, and squatters, speculators, and those with trading interests in the town of Council Grove began encroaching on Kansa land.

Naturally, the Kansa people defended themselves against the encroachment they experienced from various parties on the Santa Fe Trail and in Council Grove. New treaties were written. New promises made. Soon the railroad was talking to the white settlers at Council Grove. Demands were made that the Kansa people be expelled for the sake of "civilization." Whiskey merchants had also been part of the poisonous mix. Young Kansa men were forced to serve in the Union Army when they needed to be on the fall hunt to sustain the tribe.

Furthermore, as settlements shifting westward pushed the various tribes into each other's traditional hunting territories, new animosities between the tribes were roused. Relations deteriorated between the Kansa people and other plains tribes, and the areas in which they had hunted earlier became dangerous to them. There was pressure on the Kansa chiefs from the railroad and others looking for gain to sell their reservation, the rich land they were on, and move to a new reservation in Oklahoma. Kansa hunters returned from a hunt to find their lands occupied by squatters.

There was constant battle for control of the Neosho Valley. Tensions rose between the Kansa people and the Cheyennes. In 1868 one hundred Cheyenne warriors conducted a raid on the small Kansa tribe left at Council Grove. Inter-tribal warfare confined the Kansa people to their reservation near Council Grove; they descended further into extreme poverty. Duplicitous agents took advantage of them at every turn, joining with those who wished to force the Kansa people out of the Neosho Valley.

Eventually, a 100,000 acre reservation was offered to the Kansa people east of the Arkansas River in what is now Kay County in Oklahoma. On May 8, 1872, an act was passed to remove the tribe and sell their diminished reservation in small tracts. Chief Al-le-ga-wa-ho bitterly criticized the Union Pacific executives for their financial duplicity and demanded a fair price for his people's lands. He pled for food, clothing, schools for his children. The last forced migration of the Kansa people began on June 4, 1873, and took seventeen days as they walked south to Oklahoma.

After I had learned this Kansa history in the state named after their people, I lined up the dates with my ancestors' migration journals recording their trip into Kansas. The convergence of dates and circumstances became all too clear. Only a year after the forced removal of the Kansa people from the Neosho valley at Council Grove to Oklahoma, my own people came through Topeka looking to find, settle on, and farm land in Kansas in September 1874. My settler ancestors were shown in their first land survey sponsored by the railroad, territory for sale in the area of Council Grove, some of which must have been the very land cleared of the Kansa people to make way for "civilization," "progress," and white farmers. My ancestors chose to move on to central Kansas where they would choose their first homesteads. Yet, coming as they did so soon after the Kansa people were moved to Oklahoma from Council Grove makes clear how the process of opening tribal lands for farming worked.

Part IV

May Day

Pre-Christian May Day celebrations occurred a half-year after the November 1 mid-point between fall and winter when early blooming flowers announced the beginning of summer. The Catholic Church kept the flower rituals and celebrated the Virgin Mary. When Brown's Corner closed after my seventh grade year, and I attended eighth grade at Kismet Grade School seventeen miles east of us, I participated for the first time in a formal May Day celebration to wind the maypole with crepe paper streamers, a spring dance with roots in primitive spring festivals. Girls dressed in pastel dresses for spring. Though our maypole celebration included music and careful rehearsal and took place on a high school gym floor around a metal pole, it is clearly connected to the old Celtic or Germanic rite of spring, the holy or honored tree, the *axis mundi* reenactment of the relationship between earth and heaven.

Springtime at Brown's Corner found us training for Play Day, the track meet competition among country schools held in early May when a chill was still in the air. It was a coming out, a shedding of winter wraps, competitions sometimes not unlike initiations into age groups according to how fast a girl could run or how high she could jump. The end of school picnic, too, was a rite of spring—through my high school years. Our parents came to school for the evening picnic, and we celebrated the end of an academic year with early summer picnic food, using the occasion to say goodbye to classmates and teachers for the summer and hello to freedom from the school regimen.

Another spring celebration was our week of Bible school at church after public school was dismissed. Bible school was a celebration of nature, especially flowers, and the outdoors. We picked up our cousins and packed a full carload of children, some of whom didn't attend our church regularly; we stopped to pick buttercups from the ditches along the fourteen-mile route to our country church. We lined up outdoors on the church steps to proceed into the sanctuary to the sounds of meadowlark trills sung from their perches on the fence posts alongside our church building. We paraded into the sanctuary singing "This Is My Father's World," the anthem pounded out by the piano with a march tempo. We skipped for joy to be among church friends for the week to play and learn. No art classes were ever offered at my one-room country school, but Bible school emphasized arts and crafts projects—wood burning, painting, string crafts, fabric projects, repurposing maintenance items into art, a certain spring aesthetic we longed for.

probably 1905-1925

Chapter Nine

Homesteaders

Children, unconscious as they were of needing to use the land for anything other than delight, adapted to the new country most quickly of all and loved it all their lives.[37]

When I read accounts of the late-nineteenth century pioneer spirit of the homesteading Kansans, for example, by Craig Miner in *West of Wichita*, describing Kansas as a "garden waiting to be changed to a tame one,"[38] I realize how different our re-settlement was in 1950. Our land had been ravaged—the blood and guilt of breaking the virgin soil on someone else's hands. The 1933 tornado and the Dust Bowl winds had claimed their victories. The 1950s pioneers like my parents were chastened by what they had seen; fresh from the horrific visions of World War II Europe, theirs was not a hope to subdue the land but rather to be allowed to nest in it. I was born far too late to be one of Miner's early pioneer daughters, "little girls who played house with buffalo bones down in the grass where the breeze was cool,"[39] and yet, something in me understands and appreciates the image.

Strange and narcissistic though it may appear, as a young child who grew up in one spot and never lived anywhere else but on our familiar farmstead, though I sometimes felt some ghostly presence, mostly, mine was a solipsistic universe. The land was mine—etched into my knees, the soles of my feet, my blood and my heart. That universe included Mama, Daddy, me and my baby sister, and for awhile, my Uncle Gene, Mama's younger brother and a temporary big brother to me while he helped out on the farm when my sister was a toddler. I should have known better. I knew the old tin shed had been there when Mama and Daddy had arrived.

"Homesteading" as a concept includes notions of self-sufficiency, subsistence agriculture, isolation, a family's choice to engage in government expansion to make habitable a place previously deemed undesirable—stark contrasts to how Native peoples saw their homelands. To "homestead" as we have seen it historically entails deprivation for the sake of self-determination. The concept has evolved since President Lincoln signed into law the 1862 Homestead Act whereby settlers could claim 160 acres with a small filing fee, live on the land for five continuous years, build a residence and grow crops, and then file for a deed. A second option for the early homesteader was to purchase land from the government at $1.25 per acre after living on the land six months, building a house, and putting in crops.

By 1900 as Liberal was being settled, this Homestead Act had already doled out 80 million acres. But homesteading in Kansas, especially around Liberal, was never easy. Kansas homesteaders faced drought, scarce natural resources to burn for fuel or raw materials for building houses and barns, hard winters, sickness endured long distances from doctors, encounters with outlaws in an unsettled part of the country where there were no laws, and backbreaking work.

There were ways one could characterize my parents as "homesteaders" on our farm; they prodded back into productivity what was deemed an undesirable plot of land destroyed by tornado and Dust Bowl conditions. However, our family did not experience the same need to be self-sufficient, nor the isolation or deprivation the first homesteaders on our land must have known, even in the earliest years of the twentieth century when "late" homesteaders settled southwest Kansas.

Researching the land long after my mother had left it, I asked to see the abstract of our land. My mother ceremoniously got out her safety deposit box key, slipped it into her purse, and we headed for the bank. The receptionist brought us Mama's metal box filled with old papers and certificates. Mama carefully closed the door of the private room into which the receptionist had ushered us, looked anxiously at the little window as if she wished she might pull a shade against prying eyes. Then Mama opened the box and produced the abstract. It was long and yellow, rolled up like a Chinese scroll, and not very helpful.

But there were some names. I carefully wrote the name "Preston Elza McDonald, died December 9, 1924, of Kiowa County Kansas" and then Elza F. McDonald. And then strange and random dates: January 10, 1934. 9:00 a.m. to Mattie L. McDonald. June 27, 1934: Mattie McDonald to Elza F. and Viola McDonald. April 24, 1930, Elza Mc Donald and Viola McDonald to Lucie J. Furstenberg.

Eventually, I recognized Doc Blackmer's name: 1944-Lucie J. Furstenberg and John B. Furstenberg, 1944 to L. G. Blackmer. What I wanted to know is who built that tin shed and the old silo—essentially the only structures worth having when my folks came to the farm in 1950, the tin shed with its wonderful wooden granaries along the east side, supposedly varmint-tight wooden room-size boxes our daddy rarely filled with grain or anything else, and in whose dark and cool dusty interiors my sister and I whiled away long summer days. I will never know for certain who built it, so sketchy are the leavings of our predecessors.

Preston Elza McDonald had come "prospecting" to Kiowa County in 1886 from Indiana. Kiowa County, a couple of counties east of Seward, named for the tribe which once hunted there, is known today for the town of Greensburg, wiped out by the notorious May 2007 tornado and still being rebuilt by a community that literally decided to "go green." The parents of my childhood church friend Donna who still farms in the Oklahoma Panhandle, had come to our area from Kiowa County, so she shared her history which included the McDonalds who homesteaded our land.

Named Elza for his father, the eldest son of Preston Elza and Mattie McDonald was one year old when they came to Kiowa County. Their family history suggests that they prospered on farmland east of Mullinville despite crop failures. Like many Kansas farmers, they did what it took to stay on the land; they hauled freight, worked in construction, and helped build Mullinville.

The Census records give us scant understanding of homesteaders' lives on the land until we pair it with the conditions they endured. In the 1900 Census, eldest son Elza is fourteen years old in Kiowa County living with his parents Preston and Mattie. Five years later at nineteen, he is still listed with his parents, but sometime between 1905 and 1910 he marries and moves to the farm east of Liberal. The 1910 Census reports Elza F. McDonald, aged 24 and a farmer, is married to Viola McDonald, age 19, and they have one child. In the 1915 census Elza and Viola McDonald have three children of their own as well as two other young children living with them; in the 1920 census he and Viola have five children ranging from Edna (2 months) to Forrest (9 years), but by 1925, they are back in Kiowa County, probably because the elder Preston Elza McDonald has died in December 1924. By 1930 the whole family has relocated to Mesa County, Colorado to grow fruit, apparently having had enough of the Kansas sand.

The McDonalds no doubt struggled to farm during their tenure. In 1915 Seward County recorded a record 32 inches of rain, but in 1916 they only had 10 inches of precipitation. The Cimarron River flood was in 1914. This kind of variability in moisture from year to year was surely disastrous for homesteaders. Liberal had 53 inches of snowfall in 1911-12 but only a trace in winter 1922-23. 1924 was another drought year at barely over 13 inches of precipitation. That may have been the year the McDonalds left.

Snowfall is often of little help in providing moisture on the land in southwest Kansas as the snow blows and drifts; the moisture and protective cover the snow should offer is lost to the land. The exception proves the rule: Liberal records as historically significant a period of some 45 days from December 1911 to late January 1912 when snow covered the ground. In fact, the deepest snow recorded in Liberal was 18 inches in 1911. Normally, in our part of the country the sunshine melts snow cover almost immediately.

By 1910 when the McDonalds were farming, wheat had replaced broom corn as the crop of choice. During their time on the land the McDonalds must have endured two outstanding droughts in Seward County in 1911 and 1917, years which saw only 13 and 10 inches of rain respectively. They would have had to endure the flu epidemic of 1918; the flu ban was lifted in Liberal early January 1919. The Bozarths had come at about the same time as the McDonalds and were hard at work after 1915, putting up their silos in 1918. Those who are able to hang on through these difficult cycles survive on the land. Those who move on, leaving for others to build upon what they started, must be legion, and mostly forgotten.

1913-Present

Chapter Ten

Neighbors

Our neighbors to the east, the Bozarths, were the "real" homesteaders, well-established on the land long before my parents moved to the coyote hunters' shack a half-mile west. We didn't actually "neighbor" with the Bozarths; our circles of activities barely crossed though we lived less than a half-mile apart. Their boys, older than my sister and me, attended school and church in Liberal. We attended the little rural Mennonite church in the Oklahoma Panhandle and went to school at the nearby tiny country grade school, Brown's Corner, along the Bluebell Road north of us.

The Bozarths were long established and respected in the Liberal community. "Civilization" is generally defined as like-minded groups participating in the same schools, churches, and civic organizations. I always suspected that they saw us as migrants, poor and without lineage on the land. Their mom taught school in Liberal; ours went to the dairy barn with our dad every morning and evening. But it was probably mostly about money; we had started out with none.

Jack Bozarth, our neighbor and a second-generation farmer, and my father talked farming and sometimes helped each other, but our families lived separate lives. Jack's wife Dorothy worked in Liberal and the elder Bozarths, "Old Mr. and Mrs. Bozarth," still lived in the stately, big white farm house when I was growing up, while their children and grandchildren lived in the smaller house just west of theirs on the same farm yard. The elder Mrs. Bozarth honked her 1940s car from the far side of the small hill west of our house as she came home on the gravel road to alert whomever might be heading west into Liberal to stay in their lane. We never talked with her; we only saw her drive by.

The Bozarths were long established in Seward County, and their beautiful house and barn—the barn good enough to eventually make the Kansas Humanities Council's list of the great barns of Kansas today—surely signified to my parents every day of their early years that they were far behind the likes of the Bozarth homesteaders. Andy Bozarth, Jack's dad, born in Iowa, and his brother Frank bought their half-section of land east of Liberal in 1913. Frank later sold his interest to Andy and moved northeast of Liberal.[40]

The older Bozarths, Andy and his wife Georgia, clearly had been industrious and successful. This was confirmed by memories of their youngest daughter Lucile, whom I visited in spring 2011. Lucile, slim, graceful, and gracious, left Liberal to attend the Kansas City Art Institute in 1942 before she married her high school sweetheart, Raymond Fuller. Long retired when I spoke with her, Lucile had worked for the Board of Education in Wichita and enjoyed gardening. She loved nature and her childhood. "I always say," she smiled, "that growing up where I did when I did certainly wasn't bad, but it was terrible. . . . That is, I had a wonderful childhood, but life was very hard."[41]

I already knew from reading the Bozarth family notes in Seward County's history that Andrew and Georgia Bozarth had dug a well and built a granary in 1915 where they and their three daughters lived while they built the big house. They had brought by train Shorthorn cattle and Poland China hogs, mules and poultry to set up farming. That beautiful square two-story house we all admired when we drove past to check our land had been finished in 1916; they had put up the silos in 1918. The granary which had been their living quarters until the house was finished was purported to be the earliest grain elevator in our area.

However, Lucile, the youngest daughter of Andy and Georgia Bozarth, doesn't remember all of that family history. She was not born until 1924, a latecomer to their family who never had to use the outdoor toilet; the indoor plumbing was installed the year of her birth. Jack, the only boy in the family, whom she refers to as "Junior," was nine years older than Lucile.

Lucile was only five years old when the Depression began, and they lost "everything," Lucile says. She credits her mother with saving their family during those desperate times. A great cook whose biscuits were the family pride, Georgia Bozarth raised chickens and gardened, but most helpful to the family, sold eggs to Ideal Grocery in Liberal during those hard times in the 1930s. Lucile's predominant memory of those years was of washing eggs to help her mother; she especially recalled the time she washed 75 dozen eggs herself to take into town. She also helped in the garden and separated milk. They had hogs behind the silos. They survived.

The land blew so desperately during the '30s—the land that my parents would later farm, described by Lucile as "great land for growing broom corn"—that it covered the fence posts and wire the Bozarths had strung to pasture their cattle north and east of the big barn. So, Lucile became a herder during those years. She could earn a dollar for watching the cattle on horseback in the little strip of pasture where the hungry cows searched out grass; Lucile's job was to keep the cattle from straying.

Though the Liberal tornado of 1933 did not hit the Bozarths, Lucile remembers the day of eerie calm and the yellow sky. She also remembers Black Sunday, that infamous day in April during the Dirty Thirties. Her parents had gone to a funeral, and she and a friend were playing cards, "books," on the floor when it suddenly grew dark. She doesn't remember being frightened or worried about her parents; she simply recalls how day had turned to night.

Trying to remember other events from her childhood to share with me, Lucile brought out several of her father's journals. He kept one for every year to record his endeavors, his marketing, a farm journal detailing the year's weather, work, and production. My father had similar sketchy records penciled onto tablets, but her father's were good leather journals stamped with his name, shared among his children after his passing. I immediately spot the journal of 1949, the year of my birth. I had read before that it was bitterly cold on my birth date. Andrew Bozarth's records for January 19, 1949: "Jr. [Jack] and Walt ground grain cleaned the west end of farm for concreting" and on Thursday, January 20, the day after I entered the world: "The wind is strong and too cold to work." Lucile smiles. It must have been impossibly cold, she imagines, to be too cold to work.

And then Lucile has a story worth my trip to see her in Wichita, a story of our farm.

"Did you ever know anyone who lived on my home place?" I asked her.

"Well, yes, there was a Mr. Russell. . . ." Thus begins her story of the old bachelor who tried to farm our land, but seemed to know nothing of farming. He must have been a renter. He couldn't control the blowing sand.

I am curious. "Did you have interactions with this Mr. Russell?"

Lucile says that she really didn't know him except that for a while he had a young girl, a relative, she believes, living with him. This young girl and Lucile played together outdoors and in the cellar. She also remembers that he had a yellow rambling rose growing beside the corner of his house and allowed the girls to pick the roses. "He was nice to us and I liked him."

Ah, the cellar. . . . My parents had put a concrete slab over the cellar just north of our house to preserve it. If we had a patio, it was that 12-foot-square concrete slab over the cellar, very near the windmill to the east where we could go to get a cold drink out of the proverbial tin cup or climb up a ways to look out over the countryside, worrying our mother when she spotted us going too high and commanded us down. West of the concrete slab were the tall Chinese elms from which Daddy had hung long ropes for the board swing where we dangled in the summer heat. Oh, Lucile, I want to cover your hands with my own as we sit together at the kitchen table, knowing that you played in that cellar with the little girl who preceded my sister and me, that you walked around that house and picked those yellow roses long before we came on the scene.

The cellar memories come back to me—playing in that dank and musty dark, trying not to breathe the air, screeching at the movements of spiders, examining the shelves of canning jars. We rarely used the cellar for what it was intended—a shelter from storms. Mama was always more ready to go down into the cellar than Daddy who would sigh and say, "Oh, let me go out and take a good look at those clouds to see if they are actually that menacing" or he might check the windows to say, "It's going west of us. We don't need to go to the cellar." I remember only once when we lit the old oil lamp and sat huddled in the cellar feeling miserable because we couldn't tell what was going on above ground.

We loved to watch a storm come in, predict its direction, note the shifting winds with a wet finger in the air, gauge whose farm a couple of miles away was "getting it"—the rain we needed—or wonder whether it was hitting some area of the ground we farmed. We preferred not to go to the cellar.

My memories of the hours we spent playing house on that concrete slab include my harshest childhood memory. My sister and I were playing on the concrete patio, ages ten and six, when my father's uncle came to tell our mother that our father had been involved in a terrible accident at the corner where our mailbox stood just outside of Liberal. The driver of the car that collided with my father's grain truck was killed.

Our family never talked about what happened. I had to retrieve the newspaper account to learn what happened that Friday afternoon, June 19, 1959. The picture shows in the foreground the company car driven by Charles Townsend, "a 26-year-old oilfield service company salesman," its front half totally mangled. Charles Townsend worked for Chicago Pneumatic Tools in Liberal; his wife had gone with her young children, ages three and eighteen months, to give birth to their third child at Chickasha, Oklahoma, where both her parents and her husband's parents lived. The paper confirms what I remember hearing years afterward about the tragedy: that Daddy's truck was thrown eighty feet from the point of impact and landed upside down in the nearby field. The grain bed was torn from the truck and flung high before it came to rest right-side-up on the upside-down truck.

My childhood memories on the concrete slab have pulled me away from Lucile's story, but she remembers another incident about the moldy old cellar filled with debris when my parents moved to the farm which they cleared out to find intact concrete walls and a floor, all in decent condition.

Lucile tells me now of the time Mr. Russell came to their home, probably 1940 or 1941; Lucile was grown. Agitated, Mr. Russell asked the Bozarths if he might use their telephone to report to police the young guys from Liberal who had driven out and stolen the watermelons he had planted. Lucile was immediately worried that her own boyfriend was involved. She felt sorry for Mr. Russell as she heard him call the sheriff and report that he had shot salt pellets at the thieves, insisting that the authorities find and arrest the culprits. Mr. Russell needed those watermelons, Lucile was sure; he never raised decent crops.

Three weeks later, Mr. Russell went into our cellar and killed himself. Lucile assumed he had shot himself, wondering if he could have used the same gun that he had used to shoot those salt pellets. I wonder whether Lucile's older brother, our neighbor Jack, had told my father this sad cellar story. Maybe that is why Daddy didn't want to go into the cellar during a storm. Indeed, my sister and I always found that cellar to be creepy, but what dark, unused, moldy, spider-filled cellar is not? Or was it haunted by this sad spirit of the past?

Homeland and Homestead

"If a man owns the land, the land owns the man."
—Ralph Waldo Emerson

From the Old English "hamstede" *meaning home, town or village, the U.S. usage of the word* homestead *came to mean a plot of land adequate for the maintenance of a family, defined by the Homestead Act of 1862 as 160 acres. Recently, I visited the Homestead National Monument of America near Beatrice, Nebraska, a national historic site built on the homestead claim of Daniel Freeman, one of the first in this country to file a claim for land. The site exists to tell the stories of the homesteaders like Freeman who made two million claims on 270 million acres over the course of 123 years (1863-1976) in thirty states from Florida to Alaska. Nebraska distributed 45 percent of its land acreage to homesteaders, more than any other state.*

The Homestead Act was a revolutionary political concept envisioned first by Thomas Jefferson who thought to create an "empire for liberty" of this country; the act was eventually signed into effect by Abraham Lincoln in his wish that every poor man might have a home, and perhaps just as important, that free states would outnumber slave states. The railroad advertised a new Eden for the taking, dry Dakota land flowing with milk and honey to unsuspecting immigrants flocking to the United States from around the world. All that was required of a man, woman, ex-slave, or new immigrant was that the claimant be head of the household, at least 21 years old, and pay the $18 fee for filing a claim for 160 acres. To "prove up" a homesteader had to live on the land and build a home, make improvements to the land and farm for five years. The homesteader then received a patent from the government designating ownership signed by the U.S. President. The deal lasted through twenty-four presidents, Lincoln to Reagan.

Homesteaders came to be seen as the backbone of the country. "When tillage begins, other arts will follow," Daniel Webster believed. "The farmers, therefore, are the founders of civilization." Indeed, the Homestead Act affected nearly every aspect of life in this country. However, homesteading was not easy, as history has recorded. By 1878 John Wesley Powell, explorer and geologist, was warning Congress against settlement in the arid west beyond the 100th Meridian, a longitude west of which he believed farming as laid out by the Homestead Act would be impossible. He predicted bankruptcy and environmental desecration on what he called naked land which would not sustain farming. Indeed, 60 percent of homesteaders were unable to "prove up" and earn the deed from the federal government.

Perhaps the Homestead Act worked well enough on the east side of the 100th Meridian, that line running straight from Texas to Canada that places the Oklahoma Panhandle in the arid west. But in certain parts of the country on land like that around Liberal, lack of knowledge of the land could prove lethal. The story is told, for example of an immigrant who froze to death in a blizzard in southwest Kansas, a flyer tucked into his light linen coat advertising Kansas as the Italy of America, a verdant land of opportunity. The railroads owned the West and promoted it as a garden for farming even in areas where annual rainfall averaged less than 20 inches. Homesteading fever ran high and crackpot theories abounded—like the suggestion that "rain follows the plow;" sometimes the changeable weather itself was seductive, highly promising for a few years before it turned on an unsuspecting farmer.

Meanwhile, Native peoples watched as their ranges and ancestral areas were overrun and lines of ownership were drawn in their homelands in contradiction of their belief that where two people walk, the land cannot be owned. Native peoples' reservations were broken up by granting land allotments to individual Native Americans through the Dawes Act of 1887 which dissolved reservations and group life, giving tribal members land allotments and offering the rest of the land to settlers. The Ponca tribe lost 70 percent of its land in the land runs in Oklahoma.

Often the land allotments made to Native peoples amounted to leftover, unwanted, desert lands entirely unsuitable for farming. Furthermore, Native people frequently found the lifestyle on individual farmsteads inconsistent with tribal life; they had neither the interest nor the knowledge, not to mention the supplies, to farm. Accustomed to roving over the land, following the seasons, confinement to an acreage did not suit their lifestyle, culture or religion.. Inheritance of land as a concept was problematic among Native peoples. Children at boarding schools received allotments they could not farm. Sometimes several people received the same allotment, and it was divided into smaller and smaller parcels inadequate for even a subsistence living. Families homesteaded; tribal life was lived on homelands in villages. The concept of "homeland" as Natives thought of it, was domain, not ownership.

Today, as settlers learn the truth about the land on which they find themselves, many seek to acknowledge the land where they live as the homeland of Native peoples all these years after settler ancestors came to farm. It is estimated that 93 million of us alive today are the descendants of homesteaders. To try to learn of and acknowledge Native homelands, however, is often difficult. Oral traditions, historical memory, sacred geography, landmarks, the texture of familiarity and stories on the land which constitute the basis of homeland, are gone. What remain are the deeds to the land. Native peoples frequently lament that they have no homeland, wherever they find themselves.

1950-1978

Chapter Eleven

Farm Paths

My great-grandfather Julius Hinz, who came to this country at age nine in 1874, never drove a vehicle, even when he lived alone on the farm in his nineties. He walked several miles to town. He walked to the nearby farms of his children. He still dug wells by hand, letting himself down into the earth with a bucket system he had rigged in his late years. He needed to die with his feet still on a path he knew as he walked the miles into town, refusing rides from his children.

Sometimes when I remember our farm, I think of it as a network of crisscrossing paths like arteries of the body—the well-worn paths to the dairy barn, to feed the calves, to the alfalfa stack, to the chicken house, to the outhouse, to the tin shed, around the Osage Orange hedge, to the washline, to the windmill. I loved most of all the cow trails worn by our Holsteins. Soft, clean, sifted sand, indentations in the earth I could have walked barefoot through a three-horn patch, these trails felt like being on the right way, familiar, a protected route walking head down, one foot following the other. I had the sense that the way had been made.

It had, of course, by generations of animal and human pathfinders. I can't even say the word *path* without recalling one of the first Bible verses I was required to learn from the Psalms: "Thy word is a lamp to my feet and a light to my path." Staying on the path became symbolic.

Every summer afternoon around 4:00 I knew I was supposed to "go get the cows"—walk them home for milking from the pasture where they would be standing lazily chewing their cuds—in the remnants of a pond if there was any standing water—northwest of the dairy barn. I set out on the narrow cow path, a foot wide, marveling at its narrow width given the large girths of the four-legged Holsteins. The cows had intuitively followed the lay of the land to create the path, occasionally marked with interesting debris, like a devil's claw flipped from a cow's hoof. The red ants' mound would be well off to the side. I would have liked to walk the cow trail barefoot in the sand, but Mama demanded that we wear shoes to protect ourselves from the rattlesnakes we rarely saw.

When the cows heard my call, "Come, Bossy, Come, Bossy," in the rhythm and inflection I had learned from my parents, they raised their heads and slowly began to head south toward the dairy barn. Most fell into line obediently, the matriarch in the lead, full udders swaying. NEVER force a cow to run, my father had commanded, and disturb those udders filled with milk. Headed back to the corral, I imagined the Holsteins were thinking of the molasses-smelling grain mix my father had made for them in the barn's feed room. Like a chef for the bovine, he carefully studied the effects of his different mixes and proportions on the cows' milk production.

Those who walked this land before our family came followed game trails, buffalo or deer trails. Guides helped early European pioneers find their way across the prairie using well-known Indian trails. Ruts are still visible today where the Santa Fe trail found its way from Council Grove in Morris County in Kansas across the prairie by mule, oxen, wagon, and horse, widening the old trails into the wagon swales of what became the Santa Fe Trail. The railroad would use such trails and roads to chart its course. In Kansas, Highway 56 cuts a diagonal path across the state from southwest to northeast on the old Santa Fe Trail.

The early pages of Charles Edward Hancock's autobiography set in southwest Kansas describe an early Kansas journey using trails through open prairie following rivers, rocks, and cattle trails. Hancock's pioneer family settled in Stevens

County west of Seward; they used the Western Cattle Trail on which millions of cattle had made some fifty parallel series of deep ruts 100 feet wide "and then ran across the prairie swales, rounded the steeper hills and headed north toward Dodge City and the railroad."[42]

Heading west from Meade along Crooked Creek, a tributary of the Cimarron, they used the Jones and Plummer Trail first marked out by buffalo hunters. This trail came up from the south in Texas and from Beaver City in Oklahoma north to Dodge City. Hancock says that the big herds of cattle brought to Dodge City in those years were trailed

> up the Fort Bascom Trail, the Adobe Walls Trail and the Jones & Plummer Trail. It was a memorable sight to see herds of 2,500 and 3,000 head of these great horned cattle swinging along the trails, bellowing, fighting flies, their horns clicking together, their tongues hanging out and a great pall of dust hanging over them as they swept onward to the next watering place on the Beaver or Cimarron or Arkansas Rivers.[43]

The Adobe Walls cattle trail, a subroute of the Jones and Plummer, ran from Dodge City to Adobe Walls, Texas, actually catching a corner of Seward County as it angled south toward the Cimarron in Indian Territory (later Beaver County in the Oklahoma Panhandle). This trail had been abandoned by buffalo hunters after a raid by the Comanche chief Quanah Parker had forced them to abandon the Adobe Walls trading post, but for our visit our friend George had it marked out for us to see when we visited the Roy Smith Ranch on the Beaver River.

Then came settlers and farmers and barbed wire to block the old trails. Barbed wire had been patented in 1867, just seven years before my relatives arrived to farm and fence. Before barbed wire, boundaries were Osage orange hedgerows, thorny enough to stop cattle for sure, but time-consuming and slow for a farmer or rancher in a hurry to mark property. Our Osage Orange hedgerow on our farm's west boundary was a remnant of the post-Dust Bowl attempts to stop the blowing.

Barbed wire has its hazards. The "Big Die Up" in 1885 happened when migrating cattle tried to go south from the northern plains, could not cross barbed wire fences and died during the winter, unable to find pasture. Fence-cutting wars between farmers who fenced and ranchers who demanded free range were a fact of the 1880s. Today—though they acknowledge that fences are necessary for livestock—prairie ecologists are removing fences where they can. For aesthetic reasons. For a horizon. For studying the "lay of the land." For the sake of wildlife. Who hasn't seen a bird impaled on a barbed wire fence? Fence lines are often poorly managed over time, collecting trees and altering grassland options.

The trails I grew up on were highways. I have often assumed that my love for driving, covering the miles, is simply an inheritance of my southwest Kansas childhood. We drove to and from school, 17 miles each way, to and from church, 15 miles each way, 30 miles one way to see our aunts and uncles in the Oklahoma Panhandle on a given evening after milking 40 cows, and a couple of hundred miles to see my mother's parents in central Oklahoma.

When my parents could get no one to milk their cows (who wanted to milk on Christmas—even if we paid them?), my parents might rise at 4:00 a.m., milk the cows, drive four hours to central western Oklahoma where both my parents' extended families lived, manage a Christmas event from late morning to early afternoon with Daddy's brothers and sisters and a late afternoon into early evening Christmas supper on Mama's side, and drive home that same night for a very late milking. We were always a long ways from where we wanted to go. To get there meant a drive, usually south.

Most summer evenings my father, mother, sister, and I climbed into the pickup to "check the land," slowly driving the gravel road to look west into the setting sun's illumination of the rows of milo or corn or ensilage Daddy had planted. In this way we came to know each hill, where Daddy had found it necessary to replant some rows that particular year, where a cow had escaped the fence—the stories became landmarks.

The land accumulated stories. "There against that hill is where the nest of rabbits was. . . . Two coyotes crossed the road at this spot. . . . I saw a big flock of quail fly up right there," Daddy might say. Or, "there was a big snake along that fence-line; that snake was longer than six feet, I know!" He might lament that about on that fourth row is where his drill or planter had clogged, "You can see there where the wheat (or milo) is sparse." Farming in those years was an intimate experience with the land and filled with the stories of its encounters with the creatures for whom the land was home.

In *The Spell of the Sensuous* David Abram describes the function of place, memory, and story for the people known as Australian aborigines. So accustomed is the aborigine to using the landscape as visible reminder of story that the land literally speaks through the storyteller. If an aborigine storyteller rides in an automobile, his voice will speed up with the speed of the automobile as he speaks faster and faster moving by the landmarks whose stories pour from him.[44] In a much less direct way, and certainly without so intimate or knowledgeable a relationship with the land, the landscape speaks to me when I drive, elicits memory. Perhaps we carry an ancient need to record what we see as we move along the trail so as not to get lost.

Familiar is a word based in Latin for "household." For familiarity I looked to the west window of our dairy barn to wave at my parents standing by the window toward evening, waiting on a cow to finish letting down her milk into the attached electrical milking machine before they removed the suction cups from her udder and released her down the ramp that led south out to the "already milked" holding pen to let another cow enter her stanchion. We could judge how long it would be before our parents finished the milking by counting how many cows were in the "already milked" holding pen versus how many cows were standing near the barn's east ramp in the smaller holding pen waiting to enter the barn. Their familiar and routine paths to the water tank, from the holding pen, to the trough of alfalfa criss-crossed the cowlot.

I could draw from my memory forty years ago the boundaries and paths, label sand patterns, rust, worn boards, and grain smells. Down the barbed wire fenced lane leading to the Sudan patch from the northeast corner of the cow lot beside the stock tank I once, at about age 12, drove Daddy's red panel milk truck to show off to my visiting cousins that I could drive and tore through the fence on the corner, missing the ninety-degree turn east.

Past the alfalfa stack behind the feeding trough and silos behind the dairy barn to walk around the septic hole continuing westward, the calf shed where I taught many a baby Holstein calf just taken from its mother to suck my thumb before it learned to nurse from the rubber nipple bucket the soy smelling baby calf manna we mixed to replace the mother's milk we took to sell. In the open area beside the calf pen each spring Daddy helped us build the high jump stand and pit. Behind the trees in front of the calf barn where Mama butchered chickens, head toward the tin shed in the northwest corner of the square farmstead.

That tin shed and its environs were a second home for us as children. We floated "boats" of all sizes and shapes in front of the shed where rainwater sometimes collected in hollow places, shallow "buffalo wallows" in the sand, dug out by the wheels of farm equipment in front of the shed's entrance. Alongside the shed was a mounted gasoline tank from which Daddy pumped gas into farm vehicles while he pronounced stern warnings to us against falling in love with the smell of gasoline. (His baby brother had nearly died lying atop a vehicle sniffing gasoline, he warned us).

Just south from the shed ran the Osage Orange hedge to the road we drove west into town. Angling left or south from the tin shed across the yard was the chicken house very near the outdoor toilet from which one could follow a well-worn path to the house just past the swings and the concrete cellar patio. I blush today with the guilt of the blood on my hands from the sparrow executions we held along the wood plank top of the chicken house fence. Daddy had asked us to clean out the sparrow nests overtaking the holding pen in the barn. He had not expected that we would hold court and pass judgments on the "criminal" acts we accused the baby birds of committing, haughty judges lording it over the helpless.

There was once a large square of Bermuda grass under the wash line north of the house. When Daddy suffered back pain, he rigged a traction system to ease his pain and attached it to one of the wash line poles where he lay to relieve his back pain.

A wash house close beside the windmill and a hedge of elm trees running north of the wash house provided a backdrop for the basketball goal. There was a hydrant to wash the sand from one's feet nearby—or connect a hose, get a bucket and some soap and wash the car on Saturday afternoon while listening to Oklahoma University football games on the radio.

The drive everyone used to enter our yard lay just east of the house, set back from the big Chinese elm trees. Stepping inside the screened porch and onto the old porch boards which led into the house, you could glimpse my mother's old Maytag clothes washer and tubs in the washroom on the south end of the screened porch. Here I tried to keep my baby sister from crying while rocking her crib when Mama did the laundry, warning me against getting my fingers in the wringer as she poked at the dirty clothes with her long wooden stick.

The largest room in the house, our kitchen, was entered through the east door from the screened porch. The old scarred black and white floor tiles must have been laid by Mama and Daddy. Shiny black-checkered oil cloth, easily wiped free of food or fly "specks," rose four feet up on the kitchen walls. A gray and chrome table and chairs in front of Mama's corner cabinet pulls up the memory of my two-year-old sister given a chicken drumstick to chew on as she sat in her high chair. When my father saw that she had managed to gnaw off the tiny sliver bone at the top joint and get it stuck in her throat, her face already turning blue, my heart sank as I watched my father grab her and hold her upside down above the kitchen sink by her feet to fish that bone out of her throat with his index finger. She bled and lived.

From my parents' bedroom window, my sister, mother, and I waited for Daddy to appear during the worst blizzard ever, in 1957 when he had to leave his red panel milk truck in a snow drift and walk home. On the east screened porch Daddy and I sat together and ate dried herring out of the small cellophane packet because Mama said it stank too outrageously for us to open it inside the house.

Cicada song was always buzzing on that porch. If our relatives drove the distance from the Oklahoma Panhandle to visit on a given evening, and we weren't home, they left a bizarre display as a calling card. They went inside our house through doors never locked unless we were inside asleep, found my daddy's long underwear and stuffed it with whatever they could find to make a scarecrow form sitting on a chair or a hay bale holding a pitchfork. "Oh, no!" we would shriek returning home to the sight of a bizarre scarecrow figure dressed in our clothes awaiting our return. "We have missed Uncle Al and Aunt Esther." We loved their calling cards and never gave it a thought that they had rummaged through our drawers to create their tableau.

Because we ran a dairy, the barn was more important than the house. Begin with the broom I learned to use in the barn. There is no domestic chore I love more today than sweeping the sidewalk with a broom. In these days of blowing the grass from one's sidewalk or the snow from one's drive with a mechanical blower, I am the neighborhood dinosaur who slowly and happily sweeps debris with a corn bristled broom. I love holding a broom's sturdy wooden handle—though I grew up totally ignorant about broom corn as a crop or the process which converts the field product to the simple laborer's tool we used to sweep in the dairy barn. I certainly had no idea that I was being reared on land in an area once considered the broom corn capital of the world.

The smell of broom corn bristles, especially wet, is a fresh sweet smell, somewhere between straw and alfalfa, clean and green, akin to the smell of sweetgrass. A new broom, its bristles stiff and taut, straight and unworn, holds an erect but still flexible posture. Daddy bought good brooms—well-bristled, thick—not those cheap thin spindly ones that would not last. He eyed them carefully, grading the work like an artisan. I remember him bringing home a new broom and saying to me, "Try this one out," as if it were a new bicycle.

There is something very tired and beat down about a worn-down chewed-up-to-short-bristles broom. We always thought it indicated a lack of respect for one's work. We swept the barn's elevated concrete floor under the three bright orange metal Surge stanchions when we had finished milking our forty cows.

This was a Grade A Dairy—and it was clean. Mornings, we washed the barn floor, hosing it down, before sweeping out the water, but in the evening, we needed only to sweep out the grain the cows had pushed out of the feeding bins onto the floor along with the light hoof debris—sand or grass—they had tracked into the stanchions. Rarely did a cow urinate or defecate in the barn unless she was nervous. Mama and Daddy were, of course, bovine psychologists and prided themselves on knowing their cows' individual personalities well enough to tailor their approach and keep each cow calm and content for the duration of her milking.

I have an indelible image of my father walking up to a cow in the elevated stanchion, her udder accessible at eye level to attach the milking machine. The Holstein cow stood secured by the back gate of the stanchion on the four-feet-high clean cement. Daddy placed his long, gentle pianist's fingers on her flank so as not to startle her before he would say, "Whoa, Betsy . . . (or Pride, or Curly—each cow had her own name, assigned by Mama, to enter her into DHIA, Dairy Herd Improvement Association records). Daddy wiped the udder and teats carefully with warm, iodized water before attaching the suction cups of the Surge milkers. If Daddy were to be especially busy working fields or harvesting, Mama, assisted by her daughters when we were older, would milk all forty cows herself—maybe a ninety-minute job.

We rarely had to hobble a cow with the metal hobbles I still have hanging on my bookcase in my office. A nervous cow or one who had a clogged teat or the inflammation known as "mastitis" received tender loving care. Perhaps Daddy would have to milk her by hand and dispose of the tainted milk if she had been medicated. He leaned his head (under the striped railroad cap he always wore) softly against her side and milked by hand, pulling the teat rhythmically and smoothly into the bucket—both my father and the cow seemingly content. My parents played classical music on the radio in the barn for themselves and their cows.

Though I learned to love a broom in the milking parlor, I have never seen broom corn growing and had no idea that it was a type of sorghum from central Africa introduced to the U.S. by Ben Franklin in the early 1700s. Initially a garden crop grown for home use, by the 1860s Illinois was the leading producer of commercial broomcorn. Nor did I realize what intensive labor is required for harvesting broom corn: I have learned that today most of the broomcorn used in this country is imported from Mexico.

Liberal's broomcorn boom was short lived. Known by 1905 as the broom corn capital of the world, Liberal had, by 1910, replaced broom corn with wheat. Recently *The Hutchinson News* reprinted the following 1911 news item:

> One of the biggest wheat fields in southwest Kansas is that of the Terwilliger Ranch in Seward County. Owners raised 1,700 acres of broomcorn in 1910. Gasoline-powered engines did the plowing and cultivating. Broomcorn farmers were urged to be careful and not overdo this year. Spring 1910 price was as high as $275 a ton, with normal price $65 to 80. The high price was due to the scarcity of broomcorn in the area. Also, farmers were planting broomcorn and kafir corn where the wind had blown the wheat out.[45]

In addition to learning to wield a broom in the dairy barn, I also learned there that it was good for a girl to be strong and tall; I was praised for my ability to lift and carry two heavy buckets of the "calf manna" we mixed to feed baby calves. I credit the family business too, for helping me as a 1950s girl to understand woman strength and woman independence. My mother had helped my father dig by hand in the sand with a spade the foundation for the dairy barn they built together. Daddy hauled the sacks of cement, mixed and poured the foundation and built the structure with cinder blocks that the two of them could pick up and carry in their hands, one by one.

In 1951 with FHA loan money, and with Daddy's older brother's help, both families drove to Joplin, Missouri, to buy two truckloads of Guernsey heifers that would freshen (calve) and need to be milked in the old barn where Daddy had made stanchions. The new Grade A dairy barn would be finished in due time. Those first cows were Guernseys and Jerseys chosen for the high butterfat content of their milk. Never having been milked, of course, as these heifers freshened into milk producing animals, they entertained my parents with their own rodeo.

Unfortunately, the winters in southwest Kansas were too hard for the smaller, finer-boned Guernseys and Jerseys, and Mama and Daddy learned from dairy farmers in the area that the bigger and stronger black and white Holsteins were a better choice. These German-bred cows supported our farm until Mama and Daddy gave up the dairy shortly before Daddy died, more than a quarter century after they dug the foundation for the dairy barn by hand.

A key feature of our farmscape in the '50s was the outhouse. Mama later admitted that it was an embarrassment for her, and she worried about her daughters going out into the cold, but we took it for granted. Though there were others in our church who also still used an outhouse, most of my friends in school did not grow up with that throwback to earlier times. My parents were saving to build a new house and did not want to spend the money to remodel our little house and put in a bathroom. We referred to our outhouse as a "toilet," or, euphemistically and humorously, in recognition of our backward ways, we used the graphic old Plaut Dietsch word which sounds to me even today like the correct designation, a word pronounced "heesch-tcha" meaning something like "behind."

Growing up using an outhouse stays with you. To this day it feels somehow wrong to me to use a sparkling white toilet bowl for what belongs in the soil. Like the scat of all other animals I know, human waste belongs in the ground, and I am sometimes haunted by thinking of sewer lines running under cities everywhere. Eleanor Roosevelt would have been proud of our outdoor toilet. Indeed, Mama tells me the one I grew up with was one of those Workers Progress Administration structures for which Eleanor had campaigned in her quest for better rural sanitation. The Eleanor potty was placed on a four-foot-square concrete slab and had specifications for height: seven feet in the front, sloping a foot to its back. Carefully vented, our outhouse over the sand was not smelly. Periodically, my father moved it onto a clean new hole in the sand and buried the old waste pit.

The worn seat boards were soft and smooth, polished by the seated members of my own flesh and blood. I still like being seated on a wooden toilet seat. Complete with the requisite sack of lime in the corner to dip and scatter over one's business, we also had an always-filled toilet paper roller. I remember an old Sears catalog in the outhouse for one's reading pleasure, perhaps, or to compare one's body shape to that of the grown women models, but certainly not to use. Our two-seater helped a child mark progress as she grew, knowing when her legs were long enough to use the higher, larger seat. We had no need to measure and mark our height in inches on the wall.

An outdoor toilet also marked one's emotional or psychological progress, as in the art of mastering one's fears. Because I was afraid to go out alone in the dark, my younger sister came with me into the dark to the outhouse and stood alone outside waiting for me until I felt embarrassed by her courage where mine was lacking. We sat in our toilet alone, but my daddy's baby brother tells me that he remembers once sitting in the toilet with his mother while she pulled a sliver from his finger.

Our family has our own set of outhouse tales, in sickness, health, and celebration—like the story my mother tells of the time she and her Oklahoma Panhandle sisters had planned a surprise birthday party for my father. The plan was for the Oklahoma families to intercept our family just as we were leaving for an evening church service with the surprise birthday cake they would bring, but, driving thirty miles as they had to, they were late. My mother was determined to carry off the surprise and to my father's chagrin, she remained in the outhouse complaining that she was ill when my father called to her to hurry or they would be late for church. My uncles and aunts finally arrived to relieve my mother from her stalling duties in the outhouse.

Using an outhouse used to be a practice that marked the land dweller, put her nearer the earth and her own bodily functions, made her aware of her animal nature, akin to the cows she cleaned up after in the dairy barn, the dog who accompanied her down the path to get there, or the chickens she heard cackling nearby.

Dairy Farmers

The notion of creation care was a fact of our existence. Farming and dairying are two forms of commodification of the land, but they are also two ways to learn to love the land and its creatures, to witness and participate in the cycles of the seasons and the natural world. My parents modeled stewardship of the land and taught us reverence for life.

We were always proud of our successful family run, forty-cow dairy. The dependence upon creatures you care for and come to deeply appreciate, know by name, habit, "personality" in twice-daily milkings, is obvious. Our baby calves were sold. We did not raise them as replacements for their aging mothers, but the milk cows remained through their productive lives as part of the family farm. They were not pets, but they had the lifespan on the farm of pets. We helped them to deliver their young; we grieved their old age and passing. The whole milk they produced was greatly valued for our own use, for selling as our livelihood, but also to be sold as a specialty item to our neighbors who wanted whole milk and drove out to our farm from town to buy it by the gallon, bringing their own containers and buying it fresh from the bulk tank, grateful for an accessible place to get unpasteurized milk.

The crops one grows on the dairy farm to feed the cows, like sileage or Sudan grass, feel more like the gardening a farm family does to feed itself than marketing. Cows relished certain grains or grasses, and we took note. Most of all, we developed an appreciation for the creature, the cow, her heft, her grace maneuvering that heft, her delicate legs that hold her, the boney frame of the head, the warmth she exudes. There is a bovine warmth about cows—their steaming bodies in winter, their breath that smells like corn, their breadth!—that is seductive, a smell I experience as maternal, substantial, and dear.

1951-1961

Chapter Twelve

Country School

The one-room, one-teacher country school I attended was a relic of past times, rather emblematic of farming trends in our area. My parents had recognized early on that they could not exist on a half section of the poor land they farmed, so they had built their dairy herd as their primary support. Others in our area were either enlarging their land operations or living on their farmsteads and finding other jobs to support their families.

Land consolidation was underway; so was school consolidation. The old model of the local country school a farm kid could walk to was barely existent in Seward County, but it did exist for me my first seven grades, a throwback to an earlier time. I knew no one in my church or among my cousins in the Oklahoma Panhandle who attended a country school like I did. I considered myself a fortunate anomaly.

As an oversized five-year-old, I began first grade a year early at Brown's Corner in 1954. The story: A man drove onto our yard (people did that in the '50s, for all kinds of reasons, but especially to sell brushes, spices, or vanilla). Maybe he was one of the Wardens, our longtime civic-minded neighbors to the south. He must have been driving around taking stock of local families. He got out of his car and walked to where Mama was hanging clothes on the line. Such visitors mostly asked for my dad unless they wanted to sell Mama something.

"What is that big girl doing out there playing in the yard when she should be in school?"

Mama must have thought with embarrassment that he assumed I was truant. "That little girl is only five years old," my mama said. "She will attend school when she turns six." (There was no kindergarten at Brown's Corner).

The man continued to look at me. "Well, I would have sworn she was. . . ." (He is thinking "older" because I was always big for my age, *way* big for my age, born my mama's first child at ten pounds and four ounces and creating those stretch marks on her stomach she would thank me for at a later date.) Mama says that after the man had studied me for a while he asked, "Is there any way you would consider sending her early? We just don't have enough kids around here to save that little country school. We need fifteen kids grades one through eight. The state requires it."

His plea must have struck a note. Mama and Daddy decided to send me early. I think Mama said I was starting to read by then anyway. She had spent hours reading to me when I was her only one; she also said that I knew "Mary, Mary quite contrary/How does your garden grow?" soon after I could talk. She doesn't say it, but I know she is thinking that was the right nursery rhyme for me to have memorized. I doubt I was an easy child. Maybe Mama sent me to school early just to get me out of her hair so she could concentrate on nurturing my shy baby sister, who would have been one year old when I started school at five.

Two miles northeast of our farm as the crow flies, Brown's Corner was another experiment on land that tended not to hold people long, a ten-year attempt to try to provide a few farmers' kids east of Liberal a local school. In 1951, Alma, District #29 north of us, and Green Valley, District #11 east of us, combined to form a new District #36, on the corner of the Bluebell Road where the Browns lived. Ten years later in 1961 the school was consolidated with Kismet, seventeen miles southeast on Highway 54.

When I began, the school was still in session for only eight months to allow school children to be released early to help on the farm. But the state soon mandated a nine-month term, and the school year was extended. During my years at Brown's Corner, school children came and went, moving in to work for local farmers or other businesses and then moving on. Only one boy, John Winkler, was in my class through all seven grades until we transferred together to Kismet.

Those seven years at Brown's Corner were my halcyon days in the small white stucco school in a sage and yucca pasture. Like my church on Sundays and Wednesday nights, my tiny country school provided a comfortable and secure weekday universe where I played and learned. We burst out into the yard at recess and raced to a favorite part of the spacious playground. We dug for days in the sand building forts and caves. We concocted highways and cities and rock studded walls. There was a a makeshift softball field laid out with gunny sacks filled with sand for bases, a sturdy merry-go-round, heavy duty swings attached to a slide and teeter totter, outdoor toilets my first year, the standard equipment.

Pauline Flowers Strickland was my first, second, and third grade teacher before she became Seward County Superintendent of Schools where she continued to sign my school certificates. She was my guide to learning, and I loved her. Pauline Flowers Strickland. I loved her three names written in beautiful cursive on my report cards. She must have been an early inspiration to move me to reclaim my maiden name after twenty years of marriage so that I too could sign three names to the poems I had begun to write. I loved Mrs. Strickland with the wide gap between her front teeth, her smoker's breath, the huge round belly set between her legs like a giant beach ball when she sat with her legs of necessity widespread. I loved her huge print dresses which seemed crisp and clean—even where she tied them at the "waist" above her stomach's bulk. I would never have thought to describe her as "obese;" I didn't know the word or the concept. Somehow her size gave Mrs. Strickland stature in my eyes.

When I look today at the little black-and-white photo of her which she signed "Mrs. Strickland 1955," she seems as familiar as when I started school. At my first Christmas program before I would turn six in January I was assigned a "piece" as we called them then, as in "You will need to say your piece for your grandparents at the Christmas dinner." In my new red Christmas dress that Mama had sewn for me I was to begin the program by coming out from behind the red velvet curtain strung on a wire at the back of the large school room to create a stage for the occasion, climb up on a high stool and say,

My goodness sakes alive,
If you were only five,
Could you sit still a week
And learn a piece to speak?
I couldn't either . . . so
That's all I know!

Everyone laughed, and I suppose it was a hit, but I remember that it seemed such a simple ditty that when I sat down, I felt demeaned and embarrassed. I thought I could have memorized something more difficult. After all, I would be six years old in three weeks.

On the back of the little two-by-three-inch pinking shears-edged black-and-white school photo I still have of Mrs. Strickland is written that "my teacher forfirstsecondthird grade 37 years old." The childish, loopy cursive must be mine. I thought of her as much older. Did I ask her how old she was? I remember a strange intimacy with this teacher responsible for every student, every grade. How strange it seems now that after third grade, she simply disappeared from my life. I do not remember seeing her again.

Her grade reports suggest the country school's goals. My first-grade assessment lists under Reading: "comprehends at grade level; reads at a satisfactory speed; enjoys reading good books." Under English: "uses good vocabulary" (I started with B's and moved to A's); "speaks correctly" (B's to A's, I'm sure thanks to my parents); "good written expression;" "tries to correct errors" (again, started B, and moved to A's). Did my teacher purposely give me B's to motivate me, as I would later find myself doing with borderline cases in a college composition class?

And so it goes—social studies, arithmetic, writing, science, health, music, art and crafts, social habits, growth chart, attendance, and . . . added in the teacher's pen, a line called "conduct" where I never got above a B (even receiving two B-minuses) and about which I remember no scolding even though my parents expected A's.

In contrast, I do remember the scolding Mrs. Strickland gave Douglas Maupin and me when she kept us in at recess after she found us kissing in the cloak room and asked us whether we thought that she should inform our parents. Surely, my third grade kiss was harmless even if initiated by an older boy; this was a hazard of the close mixing of grades one through eight. Today, however, it seems highly inappropriate in a way that never occurred to me then, that a teenage boy kissed me in the cloak room. What remains in my memory is lying awake at night worrying that Mrs. Strickland would call my parents.

The grade card recorded what a doctor's office might offer today. I do not recall going to a doctor or dentist until high school. I grew three inches that first grade year. I gained five pounds. I missed only two days all year. The superintendent of schools for Seward County sent a nurse to check our teeth and tell our parents if we needed to go to a dentist; we dutifully swallowed the sugar cubes dotted with red cherry syrup, our polio vaccine. Physical education "tests" were administered each spring at the track meet.

At Brown's Corner we were on individual academic programs, each one of us in all eight grades essentially home schooled. The standardized tests we took told our parents if we were performing at grade level. On the back of the report is "The American's Creed" by Wm. Tyler Page, outlining a belief in the United States of America and a pledge to support its laws and respect its flag—that flag for which each of us kids impatiently awaited our turn to hang its furled length on the pole outside each morning and take it down each evening, without allowing it to touch the ground. We learned the appropriate etiquette for folding the U.S. flag into a tidy triangle before we laid it neatly on the teacher's desk in front of President Eisenhower's picture with its good-natured grin.

My first "boyfriend" at Brown's Corner was a beautiful Indian boy named Leonard Taylor who allowed me to use his black leather softball mitt while he was up to bat when we played workup at recess. I did my part to carefully keep that mitt from getting sand inside—the same mitt on which his little brother had spilled his mom's perfume. I could smell that mitt's lingering leathery sweetness on my hand for hours after I had used it, causing me to swoon in my earliest puppy love as I sat at my desk and worked math problems.

Leonard Taylor's parents were Roy ("Slim") and Rita Taylor, friends of my parents who lived in our community for several years. Rita was an Indian; she was also Apostolic Faith by denomination, as I recall, and an immaculately clean housekeeper. In my family faith and cleanliness were apparently far more important than race or class.

In fact, at Brown's Corner we were the children of a motley crew of individual families struggling for our own private dreams. My family was Mennonite; the Winklers were Lutherans, as we learned when Daddy went to a funeral for one of their family members and came home to report that he had enjoyed a great theological discussion with their Lutheran priest. The Arnetts were cowboys and loved horses. On occasion, I might be allowed to go home after school with Alana Arnett for a few hours to ride her horses and lick the blue icing on graham crackers her mother served for an after-school snack. The Corrells had five children, a tremendous help when we needed enough kids to play a softball game of workup at recess.

At Brown's Corner I had very little sense of "class" as grade level or class as a marker of my socio-economic status. I did not see inside the homes of most of my school mates. Though I went to school with the Corrells and the Winklers for all seven grades, I was never in either of their homes or even on their yards.

How does a school kid understand the concept of socio-economic class? I assume if I had grown up in town, I might have had some sense of class according to dress, cars, houses, high status jobs. At church we figured out who had money by their cars, the amount of land they owned, the amount they gave to the church. But our families at Brown's Corner didn't socialize, only perhaps for a school program when we were all dressed up. I have always appreciated the way that country school experience neither privileged nor discouraged me about my status.

I was junior winner and Seward County Spelling Bee Champ twice, but that was before spelling bees were an ESPN sport, so I was satisfied to have my picture—sometimes as large as four-by-six inches, hairdo thanks to my older cousin who must have encouraged me to believe I would win or why would she have done my hair?—on the front page of *The Southwest Daily Times*.

I have Mama to thank for my spelling bee wins; she used to sit up late at night and give me spelling words from the little gray book with the national Capitol's dome faintly imprinted on its cover until I got tired, turned all scowly and cranky, and started missing words, at which time she would finally say, "You are just tired. Go to bed. You'll know these tomorrow once you've slept on them." I have never wanted to go to bed at night, no matter how tired I am. I never want a day to end. And my mama has always been there to say, "You will know this tomorrow."

I do not remember lessons in geography or history or math from my years at Brown's Corner; I remember the texture of the wood floor, the tint of the green paint, the rows of cursive letters posted above the blackboard that we practiced again and again, the desktops we raised to hide behind and slammed on our fingers, the metal containers of the desks filled with pencils and papers and glue. I remember playing softball workup games at recess, practicing high jump and broad jump in the sand, the boys' forts in the sandy corners of the playground and how they dug, day after day after day, in the earth.

I remember gazing into the high blue sky as planes flew over, the outdoor toilet, and most of all the rhythms in my head as I walked through my daily regimen: the "bum bum bum" of the syllable count in my head accompanying my body's routine daily motions as I entered the school building to my mind's drumbeat, hit the bar the full width of the door which swung inward, bent to slurp one sip at the water fountain, my hand hitting the steel water knob on the right side in precisely the same movement and rhythm before the beat took me up six steps, hitting and skipping in the exact same pattern every time the wooden steps to accompany that rhythm. That comforting beat played in my head focusing me, keeping me on task, offering me my marching orders.

A long two-tiered shelf of books extended the entire length of the south wall under the sunlit windows of our classroom. We spent hours during lunch break chatting, all eight grades together, over the floor furnace, a three-by four-foot iron grate. We shoved each other to command our spot, warming our growing bodies, our legs spread wide so as not to burn our rubber soled black and white saddle oxfords, our skirts billowing as the radiating heat warmed our cold legs. I remember the chalky smell of the cloak room where we did small group work in the closet-sized drill rooms in front of both the men's and women's bathrooms where we hung our coats.

It was the 1950s and there were bomb drills during which our teacher huddled us, heads down and arms over our heads, in the dark fire exit stairs which led out to the south from the school's basement where we played ping pong. I remember the yellow and black "fallout shelter" sign on the exit door; I always wondered what the three triangle shapes might mean. I knew that there was an adult world of meanings and symbols about which I should not ask or expect to know. I remember looking up into the sky when I heard a jet go over during recess to wonder if it might be the Russians come to bomb us.

Most of all, I remember the lunches we packed in a series of lunch boxes which made their chipped and rickety ways through our lives as we carried and discarded them, each equipped with its own thermos for soup, which we rarely took. We ate sandwiches, apples, bananas, chips—few veggies. Sometimes we forgot our lunches, too hurried in the morning to pack them, or were out of what we needed to make sandwiches. On such a special day, Mama or Daddy would promise to bring us a lunch before noontime, and after the morning milking they would run into town and deliver to the school something specially "boughten" for our lunch. I loved the little cans of jellied Van Camp's Vienna sausages with the peel back aluminum tops which we ate with saltine crackers. Even today, those pallid pink processed mini-sausages take me back to my childhood; I still love the flavor of a Vienna sausage no matter that I know today that I probably do not want to know what is ground inside them!

For a year or two, the parents at Brown's Corner decided that their children needed a hot lunch program. So each day of the week a different mother, sometimes assisted by a father like mine who had a flexible noontime schedule, delivered a hot lunch. One mom was known for her chili; another for her cinnamon rolls. But that endeavor must have become too much trouble, even for the 1950s stay-at-home rural moms like mine who had lots of work on the farm helping my dad run the dairy. Our hot lunch program was soon defunct.

When spring came, knowing we would need to compete for Brown's Corner at the track meet or "Play Day," in high jump, broad jump, ball throw, and an array of races and relays, we began to train. After we had eaten our lunches, all of us would head out running east on the paved Bluebell Road, a road without shoulders on which traffic drove fast. It seems ill-advised to me today to think of that ragged line of school kids running east on the edge of a road traveled out of Liberal by farm equipment and oil company cars. No one seemed worried about it. A larger and older boy served as lead runner, his farm shoes clunking against the asphalt. At the track meet, I could race first in my own age group and then join older kids' races and relays, trying to help Brown's Corner win. We never won against the bigger schools—Blue Bell or Wideawake, but I still have my great pile of blue, red, and yellow ribbons from those track meet days, my earliest lessons in the American value of competition.

I know now that my parents worked hard to save that little country school which must have somehow meant "prosperity" and community to them. I didn't know until I checked the Seward Country School Directory in the state archives in Topeka that my father had begun to serve on the school board of Brown's Corner as treasurer my third year of attendance and remained in that role until we consolidated with Kismet.

Brown's Corner had an enrollment of twenty-one children when I began and was one of only two schools in the district classified as "one-teacher schools." We moved from an eight-month to a nine-month term in my fifth-grade year. My last two years at Brown's Corner, the teacher, Mrs. Glover, with a

B.S. degree and ten years experience, received a salary over the $4,000 mark. Mrs. DeArmond before her had seemed ancient to me with her white hair barely protecting her pink scalp and her elderly bald-headed husband who played songs for us with a wooden block on his bald head by making different mouth shapes for the notes. I kept the beautiful crocheted blue and ecru apron Mrs. DeArmond gave me from her own hope chest until it was yellowed by age before it occurred to me that it was a relic of another era I would never pull from my linen closet.

The first couple of years I attended Brown's Corner there were still around ninety children in Seward County in rural schools. When Brown's Corner consolidated in 1961, Seward County was left with two rural schools—Wideawake and Liberty, our old competitors—with seventeen and twenty students respectively. In Kansas' 1961 centennial year and Brown's Corner's last, Pauline Flowers Strickland as Seward County Superintendent was responsible for producing the annual directory to the state; she wrote in her dedication: "To the men and women of Kansas, who in the past one hundred years, have served as school board members, to the teachers and administrators and also, the pupils in our rural schools, we dedicate this directory."

For many years I hid from my academic peers my Brown's Corner one-room school experience. I knew that in some ways I had received an inferior elementary school education. Though I read prodigiously, I never got the foundational math I needed—or geography or history or art. One teacher simply could not cover all subjects for all eight grades. In fact, I learned more geography at church when they showed us on the globe the sites of Mennonite missions at harvest mission festivals. I realized that my peers had grown up with many more books in their homes and at school, educating them in ways I had missed—though my church's training in biblical literature was foundational to my later studies in literary interpretation. Our family library consisted of Bibles, biblical commentaries, Sunday school materials, a small dictionary, farm journals and magazines, newspapers, a children's encyclopedia of the Bible, and a wonderful library of piano music my parents had bought at great expense when our piano teachers recommended it.

When we did consolidate with the larger Kismet school system for my eighth-grade year, ever afterward to languish on that hour-long bus ride to and from school, my little sister, already shy going into third grade, had little experience making her way with a new group of children. She was devastated to have to get on the bus, even with her older sister, and ride to a place where she knew no one. So, perhaps our insulation and isolation at Brown's Corner set us back with social skills. I have always been grateful, however, that my school experience in Kansas taught my family how to be part of a community where people did not believe the same; tiny though that little country school was, for us it was diversity, civics, public schooling.

Brown's Corner is an apt and concrete metaphor for the dying family farm; its demise was a harbinger of things to come on the land surrounding our farm, the last gasp of a way of life that had served local citizens during the first half of the twentieth century as people like my parents built up farms on the sandy soil. Enlargement, irrigation, consolidation, urbanization became the trends of the last half of the twentieth century. Brown's Corner, resulting from the collapse of two previously attended country schools, was lucky to last a decade. It died when the local country kids were gone.

Part V

Summer Solstice

The longest day of the year, the sun at its most powerful. A summer day's stretching into forever, the hot sun penetrating my body. Endless time. A fly buzzed over my face. I didn't move. A plane droned high in the sky. Sunwarmed into a stupor, I am powerless to lift my head to look at the plane. No wind to blow away a pesky fly. Utter stillness in air whose weight you could feel.

"Summer time . . . and the livin' is easy." The archetypal summer consciousness on the farm is a communal sense of well-being—we were warm and well-fed, our cows on green pastures, danger nonexistent, sky alight with fireflies as day turned to darkness, soft lowing of cattle in the lot alongside our play, the contented cows chewing their cuds, lazily swishing tails and stomping feet to brush away flies.

An old summer solstice tradition was to build a bonfire and sit out late into the night admiring the summer sky and its stars. The farm is dark with yard lights off, the stars clear and visible. Daddy has dug a pit in the sand not far from the house and filled it with wood he needs to burn. The roaring fire sends red sparks into the heavens. Our cousins in the Oklahoma Panhandle join us after the cows are milked. Using young and limber green switches we cut from the trees and sharpen with a knife, we roast wieners and marshmallows. Long into the darkness we carry dead limbs and scrap wood, cleaning farm debris to build a raging fire, especially with tumbleweeds exploding like fireworks. When the fire becomes embers and our parents talk softly by the coals, we range over the farm playing hide and seek under the stars.

Grace, Again: for a pastor

I
The old Celtic symbol for the Holy Spirit
was the wild goose, not the dove. Imagine
this heavy-breasted bird to complete
the flat-chested Trinity. How unlike the dove
the stealth of the goose, a water bird
with an urgent honk, gathering, always
gathering the gaggle for a swim, its black neck
arched beneath a tense, inscrutable eye.

II
Catechism: I am thirteen, dangling
from a precipice. I know my heart is hard:
I have committed The Unpardonable Sin.
Nights, I cannot unwind myself, tangled
and sweaty, from the damp sheets
or my proud, hard-beating heart. I am
allowed to drive country backroads
without a license to church to seek counsel.

III
The preacher, a stern Canadian, neck
all Adam's apple, waits at his office
alongside the sanctuary. I have come,
I say, to seek my fate, though I know: even God
can't reach the hard of heart like me. But
I will take my sentence. I want the truth.
The preacher stares at me, blinks twice,
and then throws back his head, guffaws:
Don't you see, Raylene? If God had given up
on you, you wouldn't be here asking this of me?

IV
Ah, the miraculous flight of wild geese.
Not just in their alignment, the sky's
sudden vee—but the miracle of their liftoff:

their weighty bodies need no running start,
just the will to go and wings to take them
up from earth into the sky, into the passage
ethereal, beyond where we can see.

—Raylene Hinz-Penner, 2000

1903 forward

Chapter Thirteen

Church

This chapter on my church of origin, Turpin Mennonite, is the hardest chapter I write as I seek to make sense and tell the truth about settler origins. My church family are at the same time the quintessential European settlers caught in the American mythology of the doctrine of manifest destiny. They are also the reason I love the land and will spend the rest of my life trying to de-romanticize and decolonize my own life, preaching that gospel to anyone who will listen to me.

My church mentors represent "biography as theology" (as discussed by Enns and Myers, 123) to me; their teachings and examples are the codes I live by. They certainly occupied land that had formerly been the homelands of others. They believed that farming is God's chosen way and that the plow would bring rain to untenable land, the very concept that ushered in the Dust Bowl conditions whose epicenter they continue to inhabit.

However, they also survived the Dust Bowl, learned to live on the land responsibly and even today, practice a stewardship that prohibits the use of irrigation. They taught me a love of the land so deep in my soul that it naturally propels me toward empathy and responsibility toward those whose homelands we now inhabit, taught me an attitude toward the land and a faith that inspires me to tackle decolonization work. It is because of my upbringing in that country church and their modeling and teaching that I am determined to find ways to reinhabit the land responsibly wherever I find myself.

Recently, I attended the funeral of a ninety-seven-year-old farmer at my home church. As the church buried one who had been in the church nearly a century, nearly from its beginning, I listened to his life's virtues extolled: care for the land, thriftiness, self-reliance, (saving each sliver of wood from the destruction of the local elevator and hauling it home to build his own garage and barn), the creative and artful use of his woodworking skills, generosity with his family, generosity in maintaining the church, forgiveness of his errant sons, loving patience with grandchildren, in general weathering life by hanging on.

The Turpin Mennonite church is a perfect example of the westward expansion of the settler movement. They all came as farmers looking for land, a part of the long tradition of farming, the only thing they knew. They came to Oklahoma not as colonists but as opportunists and because of social conflicts where they had lived earlier. They frequently migrated not by choice but from home places that they loved.

The first Mennonite families had come to the Oklahoma Panhandle in 1903 through the Liberal train station. I recognize every name on the list of charter members; nearly every family is still represented somehow on farmland in the area. These founding families were a conglomerate of different Mennonite branches, a ragtag group you might expect to be willing to try No Man's Land—Peters' church, Krimmer Brethren, Mennonite Brethren, and General Conference Mennonites—the latter becoming the church identity in the Western District Conference which tied us into the larger Mennonite family in central Kansas and Bethel College in North Newton.

In the 1975 history written by P.M . Franz, who had been in the church for seventy-two years at that point, he recounted what his father had told his grandfather after traveling to Beaver County in 1900 to explore the possibilities for farming: it looked good in Beaver County, but it never rained out there. The grandfather replied, "No, but if people move up there to live, God will send the necessary rain." That stubborn faith saw them through the Dust Bowl.

Early on, services were conducted in homes, in German, ministers volunteering. To maintain the German language, one month of German school was held in the summer after public school ended. Between 1922 and 1926, they switched from German to English, maybe due to post-WWI pressures or maybe just acculturation. To build their first church in 1909 they collected $656 (including $274 from a parent congregation, Hoffnungsau, in central Kansas where my father's Lutheran people had become Mennonites in 1874). They built the first church (32 by 18 by 10 feet) for $625.92, an amount which included the pastors' salaries and sugar, coffee, and coal for the church.

When the church population grew enough to need a larger church, the Friedensfeld Mennonite Church found a building abandoned during the Dust Bowl in the nearby Bluebell community; its parishioners had fled the area. The congregants met in 1939 to decide how to raise the funds to buy it. Should each farmer pledge $25 per quarter of land owned or should they simply distribute pledge sheets and see what they got? They first tried pledge cards, but Mennonites have never wanted others to know how much they owned, nor could they know for sure how their crops would produce. Finally, they simply passed the offering plate and netted $465 in cash and $230 cash pledged. Land was donated for the church site. In 1940 their new building was moved to the current site at a cost of $2,000 for materials, work, and moving, including pews. The church history says, "All bills were paid with a few dollars left in the treasury at the time of dedication."

I have often asked myself why there was so little discussion of their past story of migration in a church like mine or a family like mine. Why did I know so little about their sojourn until I sought out the stories myself in middle age? I have argued that my parents were too busy as members of that post-WWII generation trying to build a life that they had no time to dwell on the past. In their attempt to account for this tendency among ancestors to leave the past behind and focus on the future, Enns and Myers offer another explanation: "During the era of Manifest Destiny, this exclusive focus on the future [seeing the past as largely irrelevant to the future] was key to underwriting settler innocence" (64). To look at the past, whether in telling the stories of WWII, as my father refused to do, or the family migration stories of coming to the U.S. for land, was to deal with traumatic events, to remember painful partings, of course, but also to have to ask questions of how the land was attained, or to dig deeper into who might have had the land before it was farmed.

Settlers like those farmers in my home congregation knew only farming and believed farming their noble calling, their gift to country, their citizenship rights. Focus on the future. Build a new life. Go forward. Work hard and see what you can do. Enns and Myers call this attitude of focusing only on the future to the exclusion of the past and its sins, making a Faustian bargain, "continually erasing the inconvenient history of colonization behind us to secure our role as noble protagonists in our story" (64). Thus, we know more about "the devised narrative of being American than that of our own family past" (64).

Yet my experience was that within my church community there was a contagious humility, joy, and appreciation of their lives on the land. All of worship was a celebration of the earth. When I read Native prayers to the earth, they seem very much like the prayers we offered to the God who made the earth. There was in all we did a recognition of our primary calling to be stewards of the earth.

Somewhere I have read that farming in an area like southwest Kansas is impossible without great support from both family and community. I have often thought about my inexperienced father's farming of the sandy soil he worked without any expertise outside of the church community. Though he was aware of the fields alongside his own in Kansas, his real standard for assessing his own achievements and success as a farmer came from those farmers in the Turpin Mennonite Church who had better land than he did. We often drove to or from church slowly over country roads past the land of church people so that Daddy could see how other farmers' crops looked. Everyone in our church farmed, it seemed, even if they lived in town. We all needed the advice and support of a farming community. Sometimes we needed their help, their equipment, their labor.

Today, at Turpin Mennonite, some of the farmers have consolidated the small family farms of decades ago into larger operations and remain on the land. After the funeral I talked to some of them. The same model sustains them. Most do not take the risk of irrigating their land in the Oklahoma Panhandle as they do not want to incur the indebtedness. They rent and farm for family or church members who have aged, moved away, or want to keep the family land but are involved in other occupations. Some appear prosperous by the looks of their equipment. Some are college educated but have gravitated away from their early "professional" jobs to the farm lifestyle they love on family land. There is still today an abundance of farm and weather talk at a church gathering as a large percentage of the church family still farms.

In 1990 Aidan McQuillan published a study on the ethnic adjustment and influence of Swedish, French Canadian, and Mennonite farmers on the Kansas prairies. McQuillan's evidence suggests that "the influence of immigrants on their American-born neighbors may have been stronger than vice versa."[46] Among McQuillan's three groups of farmers, the Mennonites "showed the least tendency to adopt the behavioral patterns of the American born. They were also the most successful group in developing farming strategies suited to the special problems of moisture supply in central Kansas and to the uncertainties of market fluctuations. American farmers who lived adjacent to them also appear to have developed a more successful, similar system of farming than those who lived adjacent to the Swedes or French Canadians."[47]

Though McQuillan's study covers earlier years, 1875-1925, Mennonites have a history of stubborn belief in their staying power on the land, and like the early Turpin Mennonite patriarch P. M. Franz's grandfather, faith that if you go there to farm, God will provide—as will their debt-free and thrifty lifestyles. Stewardship of the land was not only preached but instilled, celebrated and praised in the hymn texts composed on the core of our theology. The most-often-sung hymn of my youth from the section on "Creation" begins, "This is my Father's world, and to my list'ning ears/ all nature sings, and round me rings the music of the spheres. . . . I rest me in the thought/ of rocks and trees, of skies and seas—his hand the wonders wrought." Our praise to God was generally in gratitude for the earth: "For the plowing, sowing, reaping, silent growth while we are sleeping, future needs in earth's safekeeping, thanks be to God" (from "For the Fruit of All Creation").

In 2007, when my rural church celebrated its centennial and congregants wrote their memories for the centennial booklet, many recalled the importance of the church's harvest celebration, a Mennonite farming ceremony not entirely unlike the Plains Indian Renewal of the Earth ceremony, a 1950s Oklahoma Panhandle Mennonite version of an ancient harvest offering.

We called it Harvest Mission Festival. Our church altar was decorated with the produce, flowers, seasonal bounty the earth produced. Though every Sunday of the year the church altar displayed flowers or grasses or seasonal displays, at harvest time the adornment was especially attuned to the abundance of the earth.

During the fall, the harvest bounty began to adorn the altar table, the rich reds of milo and the straw yellow wheat heads, ears of corn in many colors, pumpkins, squash, gourds, late zinnias and marigolds, russet grasses, huge horns of plenty cascading with colorful vegetables interspersed among religious symbols often carved of wood—crosses, praying hands, pilgrims. The altar table was covered with fall's yellows and oranges, and there might be other arrangements placed around the church—shocks of corn or milo, for example, artfully tucked into corners, even baked loaves of bread to celebrate the wheat grown in the fields.

Often the Sunday of the Harvest Mission Festival was the culmination of a week of services led by a special guest speaker, a missionary home on furlough, or another preacher. We entertained such a guest as if he (often in the company of a spouse who would certainly tell the children's story and perhaps participate in other ways) were a guest from a far country come to share life lived in another world. We killed the fatted calf (in our case, hog; we served ham). We spread our favorite potluck or ethnic dishes over table after table. We prepared beautiful music. We had a festival! We adorned ourselves. Today, as autumn sets in, I go to the countryside to glean grasses from the ditches, drag home all the colors of gourds and pumpkins and squash I can find, and festoon the mantels of my home. It is not the same, however, without the community's "sacred dances."

Theology

I have long puzzled over the brand of Oklahoma Panhandle Mennonite theology I was taught at Turpin Mennonite. It was based in the literal if careful and thoughtful interpretation of Scriptures, yet it seemed not to be heavily pious. We kids were given an almost Amish license to play a bit before we settled in. There was a strong emphasis on the Second Coming, "This World Is Not My Home, I'm Just a Passing Through." But we weren't revivalists. I remember very few altar calls except for an occasional revved-up visiting preacher. Apparently, conversion wasn't the point, despite the "born again" talk. Witnessing and the Christian lifestyle were at the heart of theology: you were to live a life that would evidence to everyone around you who you were. "Let your light so shine before men that they may see your good works and glorify your Father in Heaven." Discipleship.

Because our church community was settled by a diverse set of stragglers from various Mennonite constituencies, the theology was centered in relationship. Add to that the fact that we were a long way from the centers of power in the Mennonite Church who might have told us what to think. We had our own strong sense of self-sufficiency and need for autonomy. We ardently discussed in Sunday school and catechism our personal beliefs, but mostly not in terms of orthodoxy or conversion or agreed-upon interpretations. Our practice was fundamentally Anabaptist. The key question was, "How do you read this Scripture? What is it saying to you?" Most importantly, how is this reading going to impact your lifestyle?

I always knew that church members did not agree precisely with one another on a variety of biblical interpretations, but if we respected the way they lived out their convictions, it did not matter so much. When I railed against the beliefs of someone in our congregation, my father made clear that we did not all have to believe exactly the same way, for we had been taught differently, but we did have to stay in relationship. Our group was small, fewer than 100. We could not afford to lose anyone.

This fellowship reminds me of the model for Cheyenne tribal justice that Peace Chief and Mennonite minister Lawrence Hart talked about, a restorative justice model. The tribe could not afford to lose anyone; a way had to be found for all to be allowed to make recompense, no matter how grave the sin; a path must be offered for re-entry into community, to be allowed to come back into fellowship in the tribe. This is the danger of larger congregations where institutional orthodoxy becomes preeminent and the power to enforce orthodoxy creates splinters in the fellowship. In the institutionalized church power is given to key leadership, not distributed among those in the fellowship. In contrast, there is a greater chance for all voices to be heard, empowered, valued, in a smaller group like the rural church.

Is there a foundational reason that in this country so many resist the notion of climate change? Is this rejection based in Christian doctrine? A rejection of personal responsibility in favor of an all-knowing and all-powerful deity who absolves the consumer and encourages capitalistic greed? Freed from guilt, this thinking argues that it is not for me to figure out today; I will understand it all by and by. Did Christianity convert us from the righteous "pagan" thinking that we are one with the earth? Is this earthbound concern antithetical to Christian doctrine about heaven? It has taken me decades to recognize Christianity's complicity in the thievery of the land and the commodification of the land. I worry that

Sarah Augustine is right in her claim that manifest destiny is baked into Christianity. If that is so, it will be up to Christians to disassemble that recipe.

PART VI

LAMMAS

The Anglo-Saxon "loaf mass" (Lammas) on August 1 was a celebration of the wheat harvest. "First fruits" were used to make loaves of bread to take to church as offerings. We harvested wheat in June, and though wheat heads graced our altar table in church alongside newly baked loaves illustrating the prayer we had been taught to pray, "Give us this day our daily bread," no one I knew ground their own wheat to make flour. We did put aside seed wheat from the best yield.

Wheat harvest, a time of high excitement and intense work altered our routine dramatically. We girls sank our bare feet into the accumulating wheat kernels in the bed of Daddy's grain truck while he ran his combine late into the night's darkness. We worked at night during harvest, using the combine's headlights if the wheat didn't become too damp to cut in the night air. With our cousins whose father was helping harvest our wheat, we sifted hand to hand the harvested wheat kernels, removing the chaff to chew the kernels until the gluten became elastic, marveling at our magic transformation of the wheat kernels into "chewing gum."

By August 1, the traditional Lammas date, hot though it might still be in southwest Kansas, farm kids were more than ready to go to town for new school clothes and shoes—though even a size larger than what we had worn before felt tight after a barefoot summer. New lunch buckets, notebooks, pencils, rulers, and compasses indicated a turning. Time to give up the freedom of summer for a renewed purposefulness. Fine. We were tired of our own meandering summer thoughts, aching for stimulation and interaction.

1887-the present

Chapter Fourteen

Town

What is "town" to land dwellers like us who lived only three miles from town? Neither Liberal, our trade center in the southwest corner of Seward County, nor Kismet, our smalltown public school headquarters at the east end of the county on Highway 54, ever felt like a "hometown." I always knew myself as an interloper in those places, an unfamiliar visitor. In high school I went along to "drag Main" in Kismet, Liberal, and eventually Plains, too, after another school consolidation, but I never felt at home in town. I have learned never to answer the query, "Where is your hometown?" with "Liberal," though I grew up three miles from there. I never knew anyone in Liberal. I never went to school or church with anyone in Liberal.

In my senior year of high school, another consolidation forced us to drive even farther east to our new high school midway between the smalltown rivals of Kismet and Plains on Highway 54. In 1966 I was among the seniors of the first graduating class of Southwestern Heights, the new high school built on the far eastern boundary of Seward County. As I review my public school history, Brown's Corner to Kismet to Plains, I realize that in my rural experience it would have made a lot more sense for me to talk about being from Seward County than it would for me to say, as I have all these years, that I am from Liberal.

In the 1970s when hitchhiking was an established mode of transportation for destitute college students, my Welsh graduate school buddy at the University of Kansas hitchhiked across the country's midsection beginning in Lawrence and traveling all the way to California and back—only to report upon his return that he received timely rides everywhere he traveled—except outside Liberal. There even trucks veered around him as if he were a pariah.

His story is not consistent with Liberal's founding reputation as a town named for its early generosity with water. The story goes that in the days when the end of the trail was Dodge City, those heading south from Dodge would hit a long dry zone around what is today Liberal and find themselves desperate for water. A rancher named Rogers generously dipped from his well, refusing to take payment. Grateful folk began to say things like, "Well, that is right liberal of you, Sir" later shortened to "That's Liberal," the place where water was freely given to strangers.

Today, anyone outside of Kansas who has heard of Liberal either recalls driving through town going west on Highway 54, or they have heard about the Shrove Tuesday International Pancake Race between "housewives" of Liberal and Olney, England, an event that began the same year my family came to Liberal in 1950 and put Liberal on the map.

In addition to being our trade center, Liberal was our cultural center: we went to Liberal for our weekly piano lessons and our trip to the library to check out a pile of books. We dressed up once as a family and went to the theatre on Main Street to see the movie about Franz Lizst my piano teacher assigned. Occasionally after milking we drove in to park on Main Street in front of an appliance store with multiple televisions turned on facing the street to watch the major boxing matches televised during the 1950s and 1960s. We made a holiday excursion onto Liberal's Main Street for Christmas shopping a few days before Christmas Eve.

Of course, we loved the fact that we were so close to Liberal that we could drive the three miles into town for fun—not something our Oklahoma Panhandle church community would do in the evening. Our family made trips into Liberal after milking for late suppers at the Triple A root beer joint where we got icy mugs of root beer and greasy hamburgers. Or we drove in to Fairmont's Dairy for ice cream cones hand-dipped out of five-gallon cardboard canisters. Daddy would sometimes stop at "Sherman's Station," a Mobil gas station on the east end of Liberal on the intersection of Highway 54 and the Bluebell Road run by Daddy's cousin. We were grateful during harvest to be able to haul our grain to the nearby grain elevator in Liberal, or for the convenience of hauling our milk in to Charlie Kulow's Dairy and Creamery before bulk tank days when a truck came to siphon the milk out of our bulk tank and haul it away. A trip to the grocery store was quick. The laundromat was convenient when we needed it.

However, though Liberal was our address, Route 1 Box 99, an identification of our place on the land, it was never a home town. Liberal was transactional, not a place built on real relationships like church or school. Liberal was convenient, but we weren't connected. Town offered rewards for our farm labors, but we didn't belong there and honestly, felt rather ill at ease, sometimes a bit small. I felt the distance between myself and shoppers I saw there who clearly had money to spend. I sensed my parents' dread sometimes of their need to go to the People's National Bank in Liberal, perhaps to ask for an extension on a farm loan, perhaps after a crop failure and an inability to pay for something. Liberal was a place that had power over us; it seemed best to get what you needed and get out of town.

Part VII

Autumn Equinox

During September my parents began to get ready for the extended weeks of harvest time it would take to cut silage and fill the silos for winter, a much longer, slower, and more tedious harvest than cutting wheat that Mama and Daddy did themselves, sometimes with the aid of a neighbor. Daddy changed the header on the combine for milo harvest. Cutting fall milo could extend into late fall if the weather got bad or the milo was late in ripening. Compared to the wheat harvest, everything had slowed to a crawl. The theme now was "preparation." We moved toward the old hunter-gatherer instinct to make ready for the winter when food and warmth were not readily available. Most importantly, we prepared for the winter well-being of our dairy cattle.

The church harvest festival was both literal and metaphorical, referencing both the literal harvest of crops and the biblical reminder, "The harvest is plentiful but the laborers are few" (Matt. 9:37). As autumn moved toward fall, we turned inward, insulating the house and barn against the winter cold, making sure the fields were ready to lie dormant under the cold of winter. I always had the sense that along with my new winter coat that would prepare me to face the cold wind on the plains, it was time for me to return to my labors, to prepare myself for some work I was surely being called to do.

1950s and 1960s

Chapter Fifteen

Girls on the Land

In many ways, my sister and I were isolated and insulated from mainstream culture during our 1950s and 1960s southwest Kansas Mennonite girlhoods. The openness and flexibility we experienced in attitudes about gender roles and prescriptions surely were directly connected with the needs of our family living on the land. The rural lifestyle made specific demands on a farm family, no matter whether children were girls or boys. Our primary community was the Oklahoma Mennonite church family who prized rearing children who had developed a work ethic, no matter the gender of the child.

As a 1950s girl I had internalized certain mainstream cultural norms: for example, the belief that marriage was a girl's highest priority. I realized only after marrying at age twenty, before I had graduated from college, how unconscious this marriage priority was for me; I had never seriously thought about myself as a professional. Now I was married. Who else was I? I must have chosen to become a teacher because that profession appeared to work well for married women (and our parents guided both my sister and me into the field of education which they so admired). My community, school, church, peer groups all believed a girl's success began with a good marriage. I knew I would never come back to the farm after college.

Division of labor by gender was not firmly fixed on a Mennonite farm like ours. I did not have brothers to do farm work;

when work needed to be done, it was simply all hands on deck. Also, my mother and father ran the dairy together, modeling for us the division of responsibilities as needed. My father never went to the dairy barn without my mother.

My mother's parents' model running a farm must have been an important influence on her. Her father alone controlled the purse strings in their Depression-era farm household; my parents both reacted against that model. They openly lamented my grandmother's lack of financial independence. By contrast, Mama paid the bills and kept the books. I often heard Daddy ask her for the checkbook. They were generous givers to the church, especially after harvest or for special events, and I heard them discuss and decide together how much they could afford to give. I know my father was the one who asked the bank for major loans (official "head of household" required) and was also the primary decision-maker on big equipment purchases, but never without my mother's input, and of course, he often needed her help as a driver if they went to get a new piece of equipment.

My mother may have gotten her start doing farm work because of her place as the third of three girls born before the boys in their family. The first boy, born after my mother, was nearest her age and the two of them did farm work together while the two older girls worked in the house with their mother and helped to care for the younger children.

I cannot think of a girl—whether she had brothers or not—in our church who did not drive tractor for her father, haul grain, and do farm chores. We all saw our farms as family operations, and gender seemed irrelevant when work needed to be done, especially at harvest.

This was probably especially true on our marginally arable farm. Hired help was not an option. What we four could not do, we had to ask the help of extended family or neighbors to do. Because our operation was small, if we had two trucks hauling grain as my father drove the combine, Mama always drove our truck until we girls were old enough to drive into town to the elevator, delivering harvested grain; a neighbor or relative would haul the harvested grain with another truck to keep the combine going nonstop. Then Daddy would take his truck and return the borrowed services to the neighbor who had helped him.

Probably, learning to drive very young was one way we could prove ourselves worthy contributors on the farm, and we all wanted to be mature and responsible enough that our fathers and mothers valued our contributions.

We began driving as soon as we could see over the steering wheel of the small work tractor or the pickup. First, we sat on our father's lap while he drove around the yard, carefully observing how to start, shift, brake, and manage the switches and levers. He took the time to teach us, anticipating our help when we were old enough. Then we began to drive across the yard, at the age of six or eight, believing ourselves important and helpful if our father said, "Go get the pickup and bring it here. Can you do that?" Or, Daddy might inform us that he needed to build a fence around the Sudan patch and he needed someone to drive the pickup from fencepost to fencepost as he walked, dug holes, and strung wire. We carted his equipment and learned to drive the panel truck. With a standard shift, we might kill the engine repeatedly until we got the hang of it, but we lived for our father's praise: "Ah, that was a nice smooth takeoff."

We were humiliated when he had to come restart the engine we had killed through ineptitude or had to come get us out of the sand if we got stuck. One of my life lessons was when I was plowing a sandy field north of our house and somehow got the plow clogged and twisted in field debris. I stopped the tractor and tried to dig out with my hands! Soon I saw my father come walking across the field.

He assessed the damage. I think I may have even broken something on the plow. But before we set about fixing it, the lesson. Never panic and do something stupid, my father said, like try to dig this out with your hands. (I suddenly had the image of myself as a desperate varmint trying to dig that plow out with squirrel "hands." I felt immensely foolish and inadequate). Always come get me, my father continued. You could tear up your hands digging in this field debris.

We were farm laborers, but not tomboys. Dress, makeup, and adornment were important parts of our lives. Our mother loved nice clothes, was an expert seamstress, and delighted in custom designing clothes for us for church and school. Mama loved shopping and buying good fabric and combining patterns, and she often stayed up late at night to finish new outfits we could wear to church or a special school function. Daddy too took time to admire our clothing, shoes and hair before we left for church or school, complimenting us on our appearances (and later, begging us not to wear our skirts so short). As I look back, I think girls and women in our church took pride in working as hard as the men did during the week and looking like they had not done any farm work on Sundays.

In fact, my mother worked longer hours than my father because she did her work as a dairy woman and in other ways assisting my father with farm work, and after that, all the work necessary for good housekeeping. Though Mama did "men's work" on the farm, Daddy seldom did "women's work," which is not to say that he didn't work hard. I realize now that women like my mother on the farm in the 1950s were amazingly versatile, skilled, and integral to a successful operation. After I left the farm, I read the old adage, "Man's work may be from sun to sun, but woman's work is never done," and recognized my mother's life. Mama worked late nights sewing or doing food preparation or house maintenance in addition to raising children, helping to run the dairy, gardening, preserving food, or helping us with homework. Daddy, of course, helped to raise the children. It is much easier to send a child outside to be with her father if he is doing something on the farm than if he has a job away from the home.

One very clear role for women on our farm was doing the laundry and ironing. I remember helping my mother do the wash with the old Maytag on the south end of the screened porch and hanging the clothes on the line in the early years. But when the washer finally broke down, for several years we drove into Liberal to the laundromat on the east edge of town.

Mama, my sister and I went in to do the wash. We loved this outing. Mama loved it because it was a time saver: she could take the laundry in during the afternoon and complete it in a couple of hours, stop for groceries, and easily be home for evening milking. She taught us to sort the clothing carefully and efficiently in the washers; we happily put quarters into each machine, and then we each got a bottle of pop and a magazine from the table and sat reading and commenting on the magazines the laundromat provided while the laundry was being washed. *Life* magazine was our favorite, and Mama loved to show us pictures of the British royalty, Grace Kelly, and Elizabeth Taylor. We felt somehow like sophisticated women out on the town. I greedily scavenged those magazines for news of the world until Mama called me to help fold the warm clothes tumbling out of the dryers.

We also loved going along with our daddy on errands into town. As I remember going to the Golden Gloves boxing ring with Daddy, to the male bastion of his cousin Sherman's Mobile Station, or just hanging out in our tin shed as Daddy talked with his uncles when they stopped by, I realize that one of the greatest gifts our father gave us was never to insinuate that he wished either of us had been a boy to help him on the farm. On the contrary, Daddy loved girls. It always seemed to me that he wouldn't have known what to do with a boy.

My sister and I drove tractor, milked cows, ran and jumped the high jump Daddy helped us to build each spring. I was shocked to learn from other Mennonite girls that the boys in their homes got more respect than the girls. My father was very competitive and expected us to win out over every boy in our class, whether in a foot race or a history exam; his attitude instilled confidence in us. In fact, I always had the sense that my daddy, despite his dignified masculine bearing and military demeanor, was closer to his mother and sisters than to his brothers. He seemed proud, in awe of, and totally appreciative of the wide-ranging skills and competencies of his wife.

Years after I left the farm I heard a Colorado park ranger speak on women in the west; she described a Southern Cheyenne woman warrior who took up her butcher knife to avenge her rage after the Sand Creek Massacre. The butcher knife was what a woman had, and in this case she used it. In a moment of epiphany, I thought of the butcher knife as an ironic symbol—the farm woman's primary tool and weapon. Mama butchered chickens, used her butcher knife as protection—maybe even thought of it as a weapon of self-defense.

I have a very early image of my mother's strong hands on her butcher knife, as I sat on the kitchen floor watching her sharpen the knife on the stone edge of the bean pot she had received as a wedding present, a squat gray stone crockery pot with a pink rose on its front. The pot sat high atop the refrigerator (where it still sets in her home today). Mama reached to remove the lid carefully, so as not to chip it, and with the knife in her right hand set about to create a ringing twang akin to the noise created by "playing" a woodcutter's handsaw as she smoothed the edge of her butcher knife, alternating one side and then the other, back and forth, sharpening the knife blade against the crockery edge, long ago gone gray with metallic rubbings.

My mother hated butchering chickens and so did I, but she insisted that I help her. After she had retrieved the largest fryer from the chicken house, she grabbed the chicken by its scaly yellow legs and marched herself grimly to the old elm tree near the barn where she hung the fryer from the hanging tree by its legs with the piece of wire which always dangled conveniently from a lower branch.

Upside down and helplessly flopping from its wired feet, the chicken dangled as my mother grabbed its neck to stretch it taut, then quickly hacked off its head before she retreated a few steps, allowing the chicken to flop to its death, spurting blood and reddening its flapping wings. Mama threw the head to the dog, though she noted that in the hard times of her youth, they had cooked the head and legs of the chicken.

Next, Mama dunked the bloody fryer into a bucket of boiling water and called me to the scene to help pluck feathers, a smelly job I never got used to. Nor to the smell of chicken guts on newspaper, the naked bird de-gutted on the drain board in the kitchen while Mama disposed of its innards. I fled this ghastly sight until the chicken was clean for frying—or maybe to help in the fine cleaning of an occasional missed pinfeather after Mama had singed the naked bird. What I remember most is my mother's adept hands wielding the butcher knife.

Other prairie women used the butcher knife in more dramatic, even legendary ways. My great aunt, who had come to the Oklahoma Panhandle long before my parents arrived in

1950, lived on a desolate farm in an unpainted house that looked as haunted as the stories that were told of her—the way her husband awoke to find her standing over him with a butcher knife when she was on one of her jags, for example. Today, I know that she must have suffered bipolar disorder or some other untreated mental illness of the kind that plagued lonely women whose lives got chronicled under the heading of "women and madness." I never met my great aunt, a total recluse, though she didn't live far away.

The story I heard most often was of the time the "cops" came to her house to collect on speeding tickets my great uncle or one of his sons had been issued. One glance at the lonely, unpainted farmhouse should have told those "cops" not to go to the door. My great uncle himself had alternative sleeping quarters in the round top tin shed. Supposedly, the two uniformed policemen knocked on my great aunt's door.

She opened it, brandishing a large butcher knife. Pointing her knife at the combines on her yard, she reportedly said to the officers, "There are two dead men in those combine bins, and there might be two more if you don't get off this place right now." As it was told to us, they didn't check the bins in their haste to get to their car. I thought about that story repeatedly as I grew up, making a mental note that if I needed a weapon in self-defense, the butcher knife at least could serve as a threat.

Our little four-room house had a lock on the south door, which faced the road. However, our back door, which faced east and was entered through the screened porch, had no lock. I watched every night as my mother, the last one to bed, took an old twelve-inch, tough-bladed butcher knife and stuck it snugly sideways, handle in front of the door, blade inserted into the door frame so that pushing the door from the outside would thwart an intruder. I always wondered if it would hold. So far as I know, it was never tested.

Girls in our community were the inheritors of a long tradition of women's lives on the land. The women who were our models were tough but not hard, competitive problem solvers who had known hard times long before my 1950s girlhood.

Long after I left the farm I learned the story of my favorite Sunday school teacher, a tiny wiry woman who probably did not weigh a hundred pounds. Leona lived to be 107 when she died in 2021. A woman of humor, intelligence, and grace, she was a profound influence on me. As the oldest child in her family, she had helped to save them during the Dust Bowl. Fresh out of high school with a provisional certificate to teach in the Oklahoma Panhandle during the Dirty Thirties, she brought her earnings home to help the family survive while her father took the horses to help build a bridge on Highway 83, working off the farm trying to save it for his family. Feeding themselves and their animals was their all-consuming task. They could not consider leaving the area despite one child's illness with dust pneumonia because they simply did not have the money to go. "We could not have afforded the gasoline," she told me.

Tiny Leona wore her overalls to school, hauling a five-gallon can of water for the children to drink from and wash in. She changed to her professional clothes in the outdoor toilet before the children arrived. She cleaned acres of dust out of the school and built the coal fires in winter. School was called off for a week once when the wind blew hard for seven consecutive days. Leona's family survived on bread, butter, eggs, and Leona's salary while many of their neighbors left the area. Leona never got her college degree. She later married a college-educated teacher and farmer, and though she had always hoped to finish her degree, she was needed on the farm, and after marriage she had children. She poured her heart into teaching us in Sunday school and Bible school. Women like Leona taught us to believe that farm girls had grit, were both tough and intelligent.

Part VIII

All Hallow's Eve

Agricultural peoples need a November 1 celebration, a recognition of the midway point between autumnal equinox and winter solstice, Samhain, as the Celtic festival was known, to mark the end of harvest, the start of winter, the butchering of animals for eating through the long winter months. In my earliest childhood years, when the weather became cold enough to butcher in November, we joined our uncle and aunt in the Oklahoma Panhandle for a day of butchering pigs. My memory is only of the taste of cracklings, crisp-fried and smothered with syrup and raisins and pinched up by the bite with a zwiebach.

Features of an earlier human need to mark the season are still present today: for example, the belief that winter's darkness allows the spirits of the dead to roam and haunt the living. In the Middle Ages people baked soul cakes to share with the poor, lit bonfires to scare the spirits, and the Irish made turnip lanterns to scare away spirits. In the U.S., these turnips became pumpkins and jack-o-lanterns.

Halloween or "Day of the Dead," as it is known in Mexico and other cultures south of the U.S. border, is a time to make an altar in the home with candles and flowers and to visit one's dead ancestors in the cemetery, perhaps a recognition that nothing lasts. In the U.S. it has become a time to dress in frightening and grotesque ways to beg candy or terrorize others with pranks. Such a secular celebration would seem to have little place in the ritual year of Oklahoma Mennonites; indeed, we had already observed our own non-military recognition of our dead in a service of remembrance at our country cemetery on Memorial Day.

However, always eager for some break from the sameness of our farm lives and looking for ways to recognize the seasons and socialize, we celebrated Halloween at our church with our own version of "trick or treat." Congregants laid in a stock of school supplies—pencils, rulers, tablets—to be sent to the Hopi mission schools in Arizona our church supported. Church children dressed up in homemade costumes and were driven miles around the countryside to the farm houses of parishioners to collect school supplies and enjoy the homemade treats our church family had made. In bizarre costumes, our faces disguised, the game we played was to insist that the people in our congregation guess who we were behind our costumes and masks.

Our families certainly did not spend money on costumes or buy masks to celebrate popular movie figures. Rather, we pulled on our fathers' overalls, jeans and flannel shirts and stuffed them, making crazy caricatures of farm figures and scarecrows or homely, matronly, crone-like women using masks or nylon hose pulled over our faces to distort our features. What were we celebrating with such grotesque caricatures? Of course, we loved the play of masquerade, or maybe we were objectifying what we feared, laughed at, or did not understand. The creation of an alternate self was an imaginative transformation of our growing selves into monstrous creations. Like the ancestral peoples who frightened away spirits and monsters of the dark season with terrifying faces and garb, we turned ourselves into caricatures.

No doubt our Mennonite parents wanted to keep us away from secular parties. Or maybe we honored an instinctive need to recognize the darkening earth, the dying year, the dark side, evil. And, in the normalcy of farm life routines, we longed for revelry. Our Halloween event ended with a party in the church basement where we bobbed for apples, drank hot cider, and reminded ourselves that it was fall.

And then there was Thanksgiving, a celebration of survival on the land and American opportunity. We loved the idea of the cornucopia, a symbol of abundance. As an adult I learned the basics of that old English tradition of wheat weaving and made a cornucopia of wheat straw that my mother displayed for years at Thanksgiving. We especially loved the old English harvest festival hymn by Henry Alford and knew the words by heart—words we took to heart!

> Come, ye thankful people come! Raise the song of harvest home.
> All is safely gathered in ere the winter storms begin.
> God, our maker, doth provide for our wants to be supplied.
> Come to God's own temple, come. Raise the song of harvest home.
> All the world is God's own field, fruit unto God's praise to yield;
> Wheat and tares together sown, unto joy or sorrow grown.
> First the blade and then the ear, then the full corn shall appear.
> Lord of harvest, grant that we wholesome grain and pure may be.

Today we often go to Santa Fe in New Mexico for Thanksgiving where feasts incorporate native traditions of corn and squash, pepitas, cranberries and Hatch green chilies grown and roasted in the Santa Fe sun along with the traditional turkey. The cycles of the seasons reinforce mythologies the world over, but as the mythologist Joseph Campbell has said, contemporary humanity needs a new mythology which makes the earth sacred again. Native spirituality is not a "new" mythology, but for urban dwellers who have no direct experience with land, rain, wind, or even sunshine, Native practices are sometimes a renewal of sacred experience not available in their religious practices.

Driving in the Fog, Osage County, Kansas

Decade by decade I have made my slow way
across this state, southwest to northeast—
Seward to Harvey to Shawnee County—
back up the trail moving east from Santa Fe.

In Seward County jack rabbits keep pace
alongside my moving car a half mile—involved
in the race, or the whimsy—only to suddenly
ditch the contest. Here, in these skeletal

winter forests west of Topeka, white this morning
with frozen fog, in middle age I drive blind.
Intersections east of Liberal went unheeded;
no need to stop, we could see what was coming.

We rode the bus twenty miles to school
on Highway 54 and then home again, an hour's
meditation each way to watch for any sudden rise—
tiny whirlwind, small motion of the earth in time.

Cattle made their slow way outside the miles
of barbed wire fence to stand beside the highway,
raise a curious head at our passing. We drove
fifteen miles into Oklahoma to church, and then

home, twice Sundays and again on Wednesday
nights, on sandy roads with no ditches. If you
dozed coming home in the dark, subdued by religion's
latenight spell, the sandy roadside caught

a wheel, jerked you alert and spun you back
onto the track, headed for home. But this morning
my car tunnels through rock hills exploded
to build this highway, layered buttresses

to my left and right. I could fall off the earth
into miles of ice—needle-tipped trees below
this hill's rise. Once, near this spot, a magnificent
buck launched himself off an overlook to leap

onto the hood of our speeding car. Stunned,
he wobbled his bloody way across two lanes
to die on the other side. A U-turn brought us
to the sprawled brown body, a mountain man

already poised, his long knife unsheathed, begging
for our fresh kill. Grateful for some good use, we
retreated to our bruised car to follow our dim headlights
home, glancing off the cruel surfaces of the night.

today

Chapter Sixteen

"Frontier"

I have too often mistakenly used the political term *frontier* to describe the area of southwest Kansas east of Liberal, thinking of its continuous migrations and changing populations, not recognizing my settler point of view in describing the history of the land. "Frontier," centuries ago for the front line of an army, then a borderland between countries, came to mean in North America, especially in census-taking, a settler's definition for the number of persons settled on the land per square mile. On the plains "frontier" was the demarcation between European American settlements and Native American settlements or "uninhabited" land. The concept of frontier is inseparable from the notion of "manifest destiny" and carries with it notions of privileged opportunity, "vacant" land, and westward expansion.

When I have argued that southwest Kansas continues to be a new frontier, I am referring to changes in the way the land is used, ever-changing populations, new people working in new industries. When I go back to the land where I grew up today, turn off Highway 54 heading west before I get to Liberal, take the sandy roads past the farmsteads where my country school classmates once lived, I find the farmsteads dramatically changed in ways that are different from the rural land in central Kansas, for example. Rundown, overgrown, vacant of livestock, crowded with discarded vehicles parked in the yard, these are no longer farmsteads but places to live while working elsewhere, places one can rent to live outside of town.

Even if we abandon the notion of "frontier," the High Plains area was always a land of migrating populations. After the early settlement years alternating between ranching and farming, ultimately farming held the upper hand. Natural gas discovered in the late 1920s brought a wave of people and industry followed by the Dust Bowl in the 1930s which nearly emptied the land of its inhabitants. Center-pivot irrigation over the Ogallala spurred farmers to grow corn and alfalfa which then offered opportunities in cattle-feeding and meatpacking industries. Mexican immigrants followed and Vietnamese workers too moved to Liberal to work at the packing plants that still operate today.

Growing up on the land east of Liberal has given me an appreciation for the regulatory function of good government, the federal policies a community needs when individual avarice or ignorance fouls its own nest. The Cimarron National Grassland encompassing 107,000 acres was formed when the federal government bought up the most severely eroded and wind-ravished land after the Dust Bowl destruction, taking it out of cultivation and restoring it to grassland. Likewise, today all the sections east of Liberal including the one I grew up on have been restored to grassland. Though used for farming in the 1950s and 1960s, this land in recent decades has been set aside in the government-subsidized CRP—Conservation Reserve Program. No longer farmed, it serves varied government-directed purposes: it is mowed and baled for feed for cattle in the local feedlots; it is used for pheasant-hunting reserves.

Today, the descendants of the inhabitants who roamed here 1,000 years ago and eventually moved south are migrating back as noted by Duane Bozarth, who still lives on the farm beside our land in the big old house his grandparents built. In an interview he and other Liberal residents gave the sports writer for the *Topeka Capitol Journal* some years back, they discussed how soccer had replaced football in Midwestern towns like Liberal; however, the article was actually about changes brought about by immigration and the rise of the Hispanic population in western Kansas.

Duane and the others remembered with nostalgia when Liberal was a powerhouse in football (from 1979 to 1997 Liberal played in the Class 5A state championship game eight times and won it four times).[48] The men could not hide their resentment toward the soccer-playing Hispanics who by 2006 made up 63 percent of Liberal High School. As the aircraft and oil industries left Liberal, the meat packing industries came to town, bringing with them Hispanic workers. Soccer became the sport of choice, resulting in conflicts with long-established football programs over facilities, scheduling, and fans in Liberal—as well as in the other southwest Kansas towns of Garden City and Dodge City.

Around their dining room table Duane is much less dogmatic than he is made to sound in the newspaper article. He confirms our observation that our grassland looks good. I ask if there are still problems with the land blowing. "No, it doesn't blow. That sandy portion your father fought so valiantly should never be broken up again."[49] Nearly all of the land east of Liberal as far as Light's Ranch and from the Blue Bell Road to the Oklahoma border is today a part of the government CRP program. Duane can think of only four out of the eighteen sections he knows east of Liberal that are not in the reserve program.

Duane supports himself on his retirement fund earned as a nuclear engineer before he returned to the farm, I suspect, and manages the round bale operation on neighboring CRP land to sell to nearby feedlots. He has returned to his parents' and grandparents' farmstead because there was no one else, not because he wants to live here. In fact, he remembers how desperately he wanted to leave as a boy and recounts for us how he wanted a bicycle like his friends in Liberal had. But his father knew he couldn't ride a bicycle on the loose gravel of our sandy road, and refused his request.

I haven't lived on the land near Liberal for decades. It is easy for me to take the long view of time on this land, remembering *House of Rain*, Craig Childs' search for the Chaco Canyon people who went south 1,000 years ago. I like to imagine that their descendants are among those who have come back north today to this "new frontier." Such migrations over the centuries are the normal pattern of human history.

The antelope too are moving north into our region where once "the deer and the antelope play[ed]." I spotted one as my childhood friend walked the Panhandle farmland. Her husband confirmed over breakfast that he regularly spots a herd of nearly a dozen nearby. They both disdained my delight at seeing the antelope cross the road before me—animals they see as intruders come to destroy the crops and leave their antlers in the fields to puncture tires. I realize that would not be so much a problem in the ranch area beyond Clayton, New Mexico, where I usually spot my first antelope en route to Santa Fe.

The jack rabbits have not come back to the area in the numbers we once had during the fifties and sixties. A classic photo taken near Liberal in 1935, "Western Kansas Jack Rabbit Roundup," depicts a time when people rounded up thousands of jack rabbits into a catching pen and killed them with clubs. In 1935 the *Wichita Eagle* estimated that there were eight million jack rabbits in thirty western Kansas counties, destroying the meager crops left. Farmers could not even afford the ammunition to shoot the rabbits during those desperate years; the clubbings must have been horrific, because jack rabbits are known for their shrieking, biting, and kicking.

In the fifties jack rabbits waxed and waned; they thrived some years, were far less numerous other years. One of the first times I brought my college boyfriend home during a holiday break, Daddy told us that the rabbits were overpopulated and eating the crops. My husband remembers a night hunt when my younger sister drove the pickup in our fields while he stood above the cab with a 22 rifle and shot dozens of rabbits. It sounds cruel, and yet farm life as we knew it divided all nonhuman creatures into domestic animals we were there to protect and the predators we feared might prey on our domestic animals or crops: coyotes, rabbits, snakes, mice and rats, spiders, grasshoppers, mosquitoes, ants, and flies. The common fly was a dairy man's nightmare, perching on cow manure, spreading disease, and torturing our milk cows as they stomped their feet and swung their tails in annoyance, lowering their milk production, we believed.

Clearly, there is evidence all around southwest Kansas' sandy farmland today of major demographic change in both human and animal populations, water use, types of crops grown. For example, my parents spoke often of the detestable job of picking cotton they hated as Depression era children in Oklahoma. I doubt they ever thought they would see cotton growing in southwest Kansas. In recent years, however, one sees cotton fields in southwest Kansas, especially in Kiowa County just east of us, because it takes so much less water to make a crop—one-third of what corn requires. Farmers who used to plant corn, soybeans, and wheat under irrigation now experience pumping volume or allotment issues with water use. And the low price they get for their wheat some years has encouraged some to turn to growing cotton and other crops.

Those who say that the area in Seward County where I grew up always has been and always will be frontier are probably right in the sense that there will always be new opportunities on the land. Why? Because the changing weather patterns and changing water usage opportunities make sustainability in any one industry hard. Right now, the settling of the soil into its original state of grassland seems the most sustainable.

Evolution

The Apaches are right, of course, that the land makes people live right, but only in the deep time Native People understand, the "many generations hence" time we will not live to witness. Kiowa Chief Satanta was probably right, at least about our land in Seward County; it did not want to be farmed. And yes, irrigators in this area have had to learn what the Sioux name for the aquifer,"Ogallala," evocatively suggests, "to scatter one's own." Some in these parts have been scattered from the land because of the aquifer's inability to sustain their farming practices.

"No Man's Land" was in some ways not a misnomer. Only in recent years have I recognized that there were designated "No Man's Land" buffer zones between settlers and Native peoples in many settlement areas, not just in Oklahoma. All of them reflect in their nomencla-

ture that cardinal sin of the settler, the assumption that no one inhabited the land because no one farmed there. That kind of thinking has evolved as we learn to value green space, recognize the importance of ecosystems undisturbed and re-inhabited. Our evolution as adaptable humans responding to a depleted earth and correcting our mistakes of overuse seems like the more accurate way to think about the changes I once thought of as the Liberal area's "always new frontiers."

The End

Offering

In November, after the red maize hills
were lost to stubble-stalk, and field-husks
grabbed and tore at our clothing and legs,
inside the house our father quit eating.
He came to the table, gaunt and hollow,
in an unearthly gait, like a brown vapor
settling across the carpet. We had shot
quail. What need of ours prompted this
never before prepared feast of quail?
Did we think we would hold him
in this life? That his rising spirit might
be made flesh? Or, like Jacob, had we
come to steal our blessing? Cleaned of all
feathers, the tiny legs and feeble heads
snapped off like stems, we brought him
the smooth breast-mounds, our own
innocent hearts. Bubbled gently in foaming
butter, laved and turned until golden,
the raw pink flesh was transformed into bread,
manna from the wilderness. Serving the quail
steadied our hands. It was the last time we saw him
eat—breast after tiny breast, held by his own
translucent fingers as he took and slowly chewed
the sweet meat before our grateful eyes. (1990)

My father liked to recite as a "credo" the one verse he had memorized of the poem by Sam Walter Foss (public domain):

Let me live in a house by the side of the road
Where the race of men go by—
The men who are good and the men who are bad,
As good and as bad as I.
I would not sit in the scorner's seat
Nor hurl the cynic's ban—

Let me live in a house by the side of the road
And be a friend to man.

Eternally optimistic, my father was never a cynic about either life or death. When he was diagnosed in October 1978 with a terminal disease, coronary pulmonary hypertension, he faced it as if he was farming sand: plow forward, see what comes up. Don't panic. Don't despair. He kept beside him like a weather forecast with the promise of rain tomorrow a newspaper clipping in which science seemed to offer some experimental hope for his disease, which may have been congenital or may have been environmental. He lived only three months from the time of his diagnosis.

Late December 23 we left the big Oklahoma City Baptist Hospital where Daddy's heart had stopped during the supper hour. Through the dark hours of night into Christmas Eve morning we drove our mother home over the land lit with other families' Christmas joy—land which appeared that night as eerie and strange as my mother's soft weeping in the back seat. That night's route repeated the same journey Mama and Daddy had made as a young married couple pregnant with me when they left central Oklahoma in 1948 thirty years earlier and headed west to the Oklahoma Panhandle. This time, however, it was an ending, not a beginning. We all knew that our time on the land was over.

Part IX

Christmas

The Christian winter celebration completes the farm year. Close the barn door for the longest night of the year. The land is buried under snow, frozen hard. Merry making at Christmas is possible because the land cannot be bothered. Agricultural celebrations long preceded what Christians call "Christmas." The Romans brought in the evergreen tree as a symbol of life amid winter long before Christians made it a sacred symbol in their churches to the tune of *O Tannenbaum.*

My German father of Lutheran descent trimmed the lower branches of the fresh pine tree he bought in Liberal and ceremoniously lit on fire the tiny sprigs of evergreen he had cut, wafting smoke like a priest incensing our house, performing a purification rite while Mama hustled behind with a wet towel in fear that he would burn down the house. He wanted us to smell the pine.

Christmas was our biggest celebration of the year with church and school programs, gift giving, baking peppernuts and preparing other Yuletide specialties, caroling late into the night on Christmas Eve as young people drove all over the Oklahoma Panhandle after the Christmas program singing in harmony "Silent Night," sometimes in German for the old people. "O Come All Ye Faithful." "Joy to the World." "O Little Town of Bethlehem." Sometimes those we caroled to opened the door and shouted out the name of a carol they wanted to hear. Some invited us in for hot cocoa, candy and peppernuts unless it was too late as we hustled away singing "We Wish You a Merry Christmas."

December 2020

Chapter Seventeen

Long Overdue Farewell

We pull the first cedar just as the sun is flaming over the eastern horizon; the sand winnows golden against the blue sky as it falls from the unearthed roots and catches the early morning sunlight. I am stunned by the beauty of the still December morning, our breath visible as we speak together, making plans for a workday on land we haven't walked for forty years. We know the contours of this land my sister and I both circled with the plow, though it has lain under Little Bluestem and Yellow Indian grasses now these four decades. Looking west toward Liberal, the rising sun behind me, I can see that under the thick grass, the sand is the same.

Coming here from my northeast Kansas home, I can't stop staring out over the flat horizon; I remember today the cliché I once tired of hearing: you can see for miles here. Farmstead silhouettes are visible across the horizon, a view open and energizing, like taking deep breaths, filling your lungs with pure air. My sister and I rode the combine with our father as he cut wheat on this level sandy patch, thrilled if it produced well. If the rains had come.

We brought him lunch at noon or faspa at 4:00 p.m. before we were old enough to drive tractor or truck ourselves. We made chocolate milk in a gallon jar filled with ice cubes, the milk fresh from the bulk tank in our dairy barn, mixed with real Hershey's powdered chocolate, sugar and vanilla, shook the mixture until frothy to deliver a special field treat Daddy loved, wiping his mouth with his sleeve, pulling layers of dirt from his face in those years before tractors had cabs—when a farmer ate the earth he farmed to feed himself.

Daddy has been gone four decades. So have we. The land has lain fallow in the government conservation program since our mother left here. It is strange and beautiful to be here again, our feet settling easily into the soil covered thickly in deep-rooted grasses. I want to throw my arms around my sister and freeze this moment.

But we have work to do. There must be more than fifty overgrown cedars to be removed—considered by the enforcers of the CRP program the government runs for land like ours to be fire hazards. Some of these trees are now ten feet tall with trunks a foot in diameter. How hard will it be to pull them out by the roots, load them onto a trailer and haul them to the landfill where they will be shredded? We have given ourselves just this day to do the work. Testing the equipment my sister and her husband have brought from their Nebraska farm and left here before early winter darkness the evening before, we watch now as my brother-in-law wrenches the first ten-foot cedar from the ground, yanking with the grasping shear shaped like lobster claws attached to the front of his skid steer. Bushy roots five feet in length shower the sandy soil as the tree is lifted up into the sky, sun-lit horizontally.

My younger sister and her husband decided on a whim to come pull the cedars with their own equipment rather than for us to hire someone local to do it. My husband and I jumped at the chance to come help, so here we are, an aging foursome who could not have come were we not now all retired. In fact, though I continue to pride myself on my physical strength, my fear today is that I will be unable to do my job: to kneel in the soil beside the uprooted tree and help my husband attach a heavy chain to the base of the extracted tree to be pulled over the grass to the road where it will be loaded and hauled away. My sister, always the best driver, will drive the four-wheeler to pull the trees to the road.

We see now that we have neglected this project and allowed the trees to grow here far too long. We have occasionally driven by on the roads east and south of the land, traveling south to Santa Fe, for example, but the homestead Mama sold when she left is sadly deteriorated. We tend to avoid the drive along Pine Street, dreading to see the neglected farmstead. I am grateful that the overgrown cedars have called us home.

We hadn't known how good this would feel, to stand in the December chill, wrapped in heavy work coats, to dig our hands deep into the cold, forgiving sand. Over the years we have hired others to reseed, mow, hay, or rid the land of noxious weeds. By now, the cover grass is thick and healthy, shaggy like an animal's winter coat. And then, this is no ordinary year. We have all been locked in our homes during the pandemic of 2020. Maybe that is why we came, my sister and brother-in-law hauling equipment seven hours from Nebraska to the southwest corner of Kansas. My husband and I driving across the entire state of Kansas, corner to corner, to assist. We had no idea how we had longed for the winter drive to these flatlands, longed to meet in the field and work together, socially distanced, yet feeling so close. Safe at home.

We drove in separate vehicles the previous day, my sister and her husband with their semi truck loaded with the small caterpillar known as a skid steer, a four-wheeler to pull the cedars to the road, chains, a shredder—all equipment they use daily on their farm. We joined one another to eat the Mexican food we found at a nearby food truck, our motel room door open for ventilation, frightened a bit to eat together without our masks. Outside now, feeling safe on the land, we believe it to be an easy day's work with good equipment.

Today's gift redounds from the mythical human fall from grace in Eden and the subsequent "curse" of the intrusive weed. It has forced us to get our hands into the soil. Evicted from the Garden, we have become gardeners. I kneel in the sand and dig my hands deep, remember the sand's ancient history, deposited here when the Rockies buckled and the Inland Sea moved out. What memories the land holds through time, even my own three score and ten. I am thinking too of the brevity of a human life span, how laden it is with the potential for powerful mistakes. This was, after all, Dust Bowl land torn up by the plow in the 1920s to grow wheat. Most agree today that these grasslands should never have been broken for farming. Yet, ironically, my farm life here is how I came to know this land, to love it. Farming this land is how our family survived.

Energized and joyful, the four of us work hard, amazed at how quickly we are able to pull the cedars, how physically strong our aging bodies still are, how easily the sand allows for this extraction. When our husbands load the piles of cedars onto the semi to take to the landfill near Liberal, my sister and I marvel at the quick work equipment makes of the tasks our family labored over so many years ago.

Recognizing that we are saying goodbye, we decide to dig up some small cedars to plant on her pasture land in Nebraska, a kind of remembrance of roots. We labor, laughing at how hard it is to hack out the tough rooted yucca plants she wants to haul home, how much easier to take the soft sand for her grandchildren's sandbox.

My sister will later tell me that she too realized that December day that she had left the land without saying goodbye, all of us running, almost, to escape the devastation of our father's premature death. She, who has lived on farmland in Nebraska since she left, will tell me how, as she sat on the Bozarth farm where we parked our equipment overnight to look across at our land, she felt the keen familiarity in the landscape, gratitude for her years on the land. She remarked on recognizable landmarks as we dug the cedars—the silhouette against the western sky of the still-undisturbed winch on the silo Daddy designed to move the silage unloader from one silo to the other. She turned to me, smiling, "Do you remember how great it was that this land was so flat we could see the school bus coming two and a half miles away to give us enough warning to get our hair combed and clothes on before it arrived?" Later, she will put together pictures from this day to show our mother, but our sheer exhilaration in this experience will exist as a memory only the four of us will always share.

Too soon we have loaded the equipment, brushed the sand out of our clothes and headed north and east back to our homes and the lives we have made since we left here. Will we return?

Why? Under what circumstances? And what will finally happen to this land when we are gone?

I think about that sometimes. What is the right way to honor the land and our time here when we sell it? With whom should we share the proceeds of our rich lives on this land? The land itself will remain the record of our being here as it holds the stories of our wanderings, evictions, relocations, settlements, resettlements—so many human footsteps before and after ours. Long beyond the time we will all cease in our wanderings.

Notes

1. Groves, "Smithsonian Confirms," 1.
2. Everhart, *Oceans of Kansas*, 5.
3. *Ibid.*
4. Buchanan, 53.
5. Abram, *The Spell of the Sensuous*, 232.
6. Abram, 233.
7. Abram, 235.
8. Egan, *The Worst Hard Time*, 9.
9. Egan, 8.
10. Egan, 10.
11. Egan, 28.
12. Egan, 44.
13. Lorca.
14. *Ibid.*
15. Childs, *House of Rain*, 11.
16. Childs, *House of Rain*, 20.
17. *Seward County Kansas*, 38.
18. Gilbert and Brooks, *From Mounds to Mammoths*, 42.
19. Gilbert and Brooks, *From Mounds to Mammoths*, 43.
20. Schneider, "The Roy Smith Site," 125.
21. Holmes and Gilbert, "Prehistoric Panhandle Farmers: The Roy Smith Site," n.p.
22. *Ibid.*
23. Chrisman, *The Call of the High Plains*, 44.
24. Schneider, "The Roy Smith Site," 125.
25. Enns and Myers, *Healing Haunted Histories.*
26. Mann, *1491*, 251.
27. Strate, *Up from the Prairie*, 106.
28. Staples, "'On Civilizing the Nogai."
29. *Ibid.*
30. Hamalainen, 25.

31. Quoted in Gwynne, *Empire of the Summer Moon*, 154.
32. Phillips, "History and Lore of Alibates Flint."
33. Gwynne, 32.
34. Knab, *Polish Customs, Traditions and Folklore*, 91-92.
35. Saul, 38.
36. Friesen, Johann "Hans" Ediger, 83.
37. Miner, *West of Wichita*, 50.
38. *Ibid.*
39. Miner, *West of Wichita*, 51.
40. *Seward County Kansas*, 225-226.
41. Fuller, Interview, March 23, 2011.
42. Chrisman, *The Call of the High Plains*, 9.
43. Chrisman, 15.
44. Abram, *The Spell of the Sensuous*, 172-175.
45. Kilbourne, *Hutchinson News.*
46. McQuillan, 200.
47. *Ibid.*
48. Corcoran, "Rise of Soccer," 1D.
49. Bozarth, Interview, July 30, 2010.

(Several MLA style parenthetical footnotes are in the text and not included in this list)

Bibliography

Abram, David. *The Spell of the Sensuous: Perception and Language in a More-Than-Human World*. New York: Vintage, 1996.

Biss, Eula. *Notes from No Man's Land*. Minneapolis: Gray Wolf Press, 2018.

Bozarth, Duane. Interview by Raylene Hinz-Penner. (July 30, 2010).

Buchanan, Rex, ed. *Kansas Geology: An Intro to Landscapes, Rocks, Mineral, and Fossils*. Lawrence, Kan.: University Press of Kansas, 1984.

Campbell, Joseph. *The Power of Myth*. New York: Doubleday, 1988.

Childs, Craig. *House of Rain*. New York: Little, Brown, 2006.

Chrisman, Charles and Harry E. Hancock. *The Call of the High Plains: The Autobiography of Charles E. Hancock*. Denver: Maverick Publications, 1989.

Corcoran, Tully. "Rise of Soccer, Fall of Football." *Topeka Capitol Journal*, August 17, 2008: 1D, 8D.

Dismantling the Doctrine of Discovery Toolkit. MCC, 2018.

Durban, Eric. *National Public Radio*. April 11, 2011. http:m.npr./news/Business/13534631 (accessed July 1, 2013).

Egan, Timothy. *The Worst Hard Time*. Boston: Houghton Mifflin, 2006.

Enns, Elaine and Ched Myers. *Healing Haunted Histories: A Settler Discipleship of Decolonization*. Eugene, Ore.: Cascade Books, 2021.

Everhart, Michael J. *Oceans of Kansas: A Natural History of the Western Interior Sea*. Bloomington, Ind.: Indiana University Press, 2005.

Friesen, C. T. *Johann "Hans" Ediger: 1775-1994*. Hillsboro, Kan.: Multi Business Press, 1994.

Fuller, Lucile Bozarth. Interview by Raylene Hinz-Penner. (March 23, 2011).

Gilbert, Claudette Marie and Brooks, Robert L. *From Mounds to Mammoths: A Field Guide to Oklahoma Prehistory*. Norman: University of Oklahoma Press, 2000.

Groves, Esther. "Smithsonian Confirms Coronado Bit, Buckles." *The Southwest Daily Times*, July 25, 1979: 1.

Gwynne, S. C. *Empire of the Summer Moon*. New York: Scribner, 2010.

Hamalainen, Pekka. *The Comanche Empire*. New Haven: Yale University Press, 2008.

Hinz-Penner, Raylene. *Searching for Sacred Ground*. Telford, Pa.: Cascadia Publishing House, 2007.

Hoedel, Cindy. "Glory Days." *The Kansas City Star Magazine*, October 10, 2010: 7-12.

Holmes, Mary Ann and Claudette Marie Gilbert. *Prehistoric Panhandle Farmers: The Roy Smith Site*. Prehistoric People of Oklahoma No. 3, Norman: Oklahoma Archaeological Survey and The University of Oklahoma Stovall Museum of Science and History, 1979.

Kilbourne, Clara. *The Hutchinson News*. http: www.hutchnews.comj/print/MON—Yesterday-5-9—1 (accessed May 19, 2011).

Knab, Sophie Hodorowicz. *Polish Customs, Traditions and Folklore*. Hippocrene, 1996.

Lorca, Garcia. "Theory and Play of the Duende." *Poetry in Translation*. 2007 . https:www.poetryintranslation.com/

PIPBR/Spanish/LorcaDuende.htm (accessed July 1, 2013).

Mann, Charles C. *1491*. New York: Alfred A. Knopf, 2005.

McQuillan, D. Aidan. *Prevailing Over Time: Ethnic Adjustment on the Kansas Prairies, 1875-1925*. Lincoln, Neb. and London: University ofNebraska Press, 1990.

Miner, Craig. *West of Wichita: Settling the High Plains of Kansas, 1865-1890*. Lawrence, Kan.: University Press of Kansas, 1976.

Phillips, Wes. *History and Lore of Alibates Flint*. http:www.panhandlenation.com/history/alibates2.html. (accessed August 23, 2012).

Saul, Norman E. "The Migration of the Russian Germans to Kansas." *Kansas Historical Quarterly*, 1974: 38-62.

Schneider, Fred E. "The Roy Smith Site, Bv-14, Beaver County, Oklahoma." *Bulletin of the Oklahoma Anthropological Society*, 1969: 119-179.

Seward County Kansas. Liberal Kan.: K.C. Printers, 1979.

Shortridge, James R. *Peopling the Plains: Who Settled Where in Frontier Kansas*. Lawrence, Kan.: University Press of Kansas, 1995.

Staples, John R. *'On Civilizing the Nogais': Mennonite-Nogai Economic Relations, 1825-1860*. April 2000. https:www.goshen.edu/mqr/pastissues/apr00staples.html. (accessed July 1, 2013).

Strate, David K. "Satanta: Orator of the Plains." In *Up from the Prairie*, 106. Dodge City Kan.: Cultural Heritage and Arts Center, 1974.

The Southwest Daily Times. "Liberal Man Dies in Car-Truck Accident." June 20, 1959: 1.

"Timeline of the American Bison," www.fws.gov, n.d.

The Author

Raylene Hinz-Penner was born in Liberal, Kansas, and grew up on a dairy farm three miles east of Liberal on sandy Dust Bowl land she came to love. She attended Bethel College in North Newton, Kansas, and earned advanced degrees at Kansas University in Lawrence and Wichita State University.

A retired college teacher of literature and creative writing at Bethel College and Washburn University in Topeka, Kansas, she is interested in the deep history of the places where she has lived and the peoples who have inhabited those areas or migrated over the land. She wrestles with her own European Mennonite family's 500 years of farming migrations and settlement history which often forced Indigenous peoples off the lands.

Hinz-Penner lives today in central Kansas with her husband Doug Penner. In retirement she continues to write about geographic location and place, land acknowledgement, and migration. She is a member of Bethel College Mennonite Church and retains a wider fellowship membership in Southern Hills Mennonite in Topeka.

CPSIA information can be obtained
at www.ICGtesting.com
Printed in the USA
LVHW041253030523
745973LV00003B/216